Breeding & Training a Horse or Pony

Breeding & Training a Horse or Pony

Anne Sutcliffe

David & Charles

Newton Abbot   London

Illustrations by Christine Theobald

British Library Cataloguing in Publication Data
Sutcliffe, Anne
  Breeding and training a horse or pony
  1. Horses   2. Ponies
  I. Title
  636.1   SF285
  ISBN 0-7153-7953-4

Typeset by ABM Typographics Limited, Hull
and printed in Great Britain
by Redwood Burn, Trowbridge, Wilts.
for David & Charles (Publishers) Limited
Brunel House Newton Abbot Devon

Contents

I
First Considerations

Possibly the best advice that any would-be horse or pony breeder can be given is summed up in two short words: *think first* and think very seriously about what is involved. You need to give careful consideration to all aspects of the matter, working honestly through the answers to such questions as: what type of horse or pony am I trying to breed? What will it be capable of doing when it matures? Will there be anyone at home to ride it or am I breeding to sell? What will it cost to rear properly to a saleable age, or alternatively to the age when it can start work? You have to remember that should you be thinking of keeping your foal to ride yourself, it will be a minimum of five years from the day the mare is covered before your youngster is any use as a riding horse, and an awesome amount can happen to one's life in five years!

Think honestly too about facilities. Have you a safe field for a mare and foal, preferably—and essentially when the foal is very young—a small paddock for their exclusive use that can be rested for three months before the foal's birth? This means that the youngster has the benefit of relatively worm-free grass and the mare a good pull of fresh spring grazing to help her milk supply. Do you have a suitable box for foaling a mare? It need not be elaborate but it must be safe, large (at least 12 x 14 ft), dry, warm and equipped with electric light, a sound floor and a top door. Have you suitable accommodation for the foal at weaning time and for the first winter?

Give careful thought to all these questions and then if you are honestly happy with the answers, and I do stress honestly, you may think further. It is no good pretending to yourself at this point that a problem you have discovered will go away; for example, had you forgotten that your strong teenage son might well be at college by the time your three-year-old needs backing?

Are you hoping that the money will be somehow available to renew the awful fencing in the little paddock? If you cannot be certain that you will manage without your son's help, or see a definite way of sorting out a snag, better forget it. Breeding is an expensive hobby and if you cannot do it properly, cannot handle your youngster, or do not have necessary basic facilities, your longed-for foal could end up a disaster and also a vast financial drain.

Is the Mare Right?

Having thought about the general problems it is now time to be more specific. Considering the matter closely, and again strictly honestly, is the mare you have worth breeding from? Are her conformation, action and temperament sufficiently good to allow her to breed useful progeny with a future? Any unsoundness that is not a direct result of accidental injury

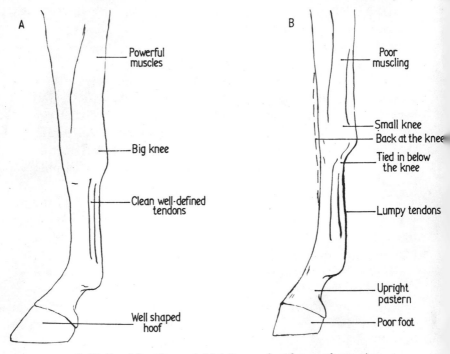

(left) Good fore-leg and (right) poor fore-leg conformation

should debar her from being considered as a brood mare, particularly if the trouble is one that is likely to be passed on to her offspring.

Faulty action resulting in constant over-reaching, wind defects, lameness caused by ill-shaped feet and poor front-leg conformation leading to recurring leg problems, are just a few things that ought to make one reject a mare as breeding stock. Study your potential brood mare and decide in your own mind what her conformation is like, what are her faults and strong points, whether the faults outnumber the good points; if they do, again forget her as a brood mare.

Remember the stallion can only pass 50 per cent of his genetic makeup to his progeny. If everything is wrong with the mare, the stallion, no matter how prepotent, can only improve a maximum of half of the faults, which may not make for a very beautiful foal.

The mare must also have a good temperament and have proved herself a reliable performer under saddle. No mare that has been shown to have a flighty, silly temperament or to be evil-tempered should ever be bred from. It makes me furious to hear folks say 'Oh well, she was so silly in traffic and generally unpredictable that we've decided to keep her for breeding from now on.' It is my firm belief that anyone who breeds from a mare that has been frightening and unreliable to ride, or unpleasant and difficult to handle, is playing straight into the hands of the meat man. Slaughter is often the only outlet for some of the unsound, misshapen young animals, many with peculiar temperaments, that are being bred without due thought as to their likely future.

No, be honest with yourself and be sure that your mare's conformation, action, temperament and ability are sufficiently good to qualify her as a brood mare before you start considering what sort of foal you are trying to breed.

Size

These days size seems to be the great criterion, but should you decide to use that 17-hand thoroughbred sire, remember—

particularly if the mare is 15 hands plus and with some thoroughbred blood—that the offspring could easily be over 16 hands by the time it is two years old, and young horses can be playful, strong and, let's face it, at times just plain sinful. Consider whether you, on your own, almost certainly without help (crises always arise when everyone is out), are capable of controlling, handling and if needs be disciplining a stroppy two-year-old colt of this size.

For myself I am a great believer in the old adage 'Like to like begets like', and so if you have the children's outgrown hunter pony aim to breed another hunter pony for someone else's child to enjoy. If you have a hunter, use a hunter sire, if a show pony use a show-pony sire. The possible exception to this is with native ponies, when an Arab sire may well produce offspring which, if a filly, will make a first class broodmare to top cross with a thoroughbred or Anglo-Arab, to breed your 15.2hh–15.3hh riding horse.

Do not be persuaded into trying to breed something bigger from the children's outgrown favourite for them to grow into. Unless you have a very 'long' family this is unlikely to work. By the time the foal is mature the proposed rider will have grown too big, lost interest, gone to college, got a job or a wife/husband, or both; likewise do not expect your flighty TB mare to breed something suitable for Grandpa to ride.

Breeding a Horse to Enjoy

In all things aim to breed an animal you can enjoy handling and having about. Breeding horses is supposed to be a pleasurable hobby; there is no point in breeding something with which you cannot cope. Either the animal will end up neurotic or bad-tempered, or you will. Do not be swayed by other people's opinions on the sort of horse you should try to breed. If you have really decided it will be best for you to use an Arab stallion and not the local hunter sire, stick to your decision. After all, it is you who know what you wish to use the foal for.

If you have thought over these things carefully and have no niggling doubts, no firmly suppressed fears about handling your

'baby horse', and a clear mental picture of both your own and your mare's limitations and the hoped-for final product, you are probably one of the few amateur breeders who will make the grade. You could have the satisfaction in five years' time of saying with pride 'I bred that' as a quality, well-mannered young horse makes its debut.

There is possibly one other note of caution. Many young wives decide to breed a foal from their mare whilst they themselves are pregnant. This does mean that the new mum is going to have two young things to care for at the same time, both of whom are very demanding on time and energy, the human variety particularly. If there is no help available to give assistance with the mare and foal it is inevitable that the education and handling of the foal will be neglected when the human mum finds—perhaps to her surprise—that her own child's welfare occupies most of the day. There is also the practical aspect of leading a yearling out whilst pushing a pram with the other hand. It is one of those good ideas in theory that frequently does not work out in practice. I am also slightly wary of very young unpredictable horses getting too involved with very young unstable humans, even though I confess we did it, with no harm to either. If you are going to mix young equines with young humans, the selection of the equines for temperament obviously becomes even more important.

One final thing that could save you a deal of time, money and worry: before the mare finally departs to stud get a well-recommended *horse vet* to examine your mare and see that she is in fact capable of breeding and in a fit condition to do so

2
Choosing
Stud and Stallion

If you do decide to go ahead and breed a foal from your mare, the next major step is to choose your stud and the stallion. I mention the stud first because in my opinion it is almost the more important. Remember, your mare will be there for upwards of a month and if during that time she is mishandled, neglected, underfed or frightened, firstly her visit to the stallion is unlikely to be successful, and secondly you will have the problems to sort out when she returns home to you.

Possibly the best way of choosing the stud is by personal recommendation. If someone you know and can trust regularly sends their mare to a particular stud and remains satisfied, then you should have no problems. It is worth mentioning that it is not always the fashionable stallions who either sire the best foals or stand at the best-run studs. To face facts, a fashionable stallion may be an excellent sire with a proven performance record, standing at a well-managed establishment and producing top-quality progeny; but on the other hand, many popular hunter sires are little else but failed racehorses with nothing but a long pedigree to recommend them. Likewise strings of show championships do not guarantee either a beautiful foal or the ability to perform well under saddle.

So think carefully before you patronise the local stallion of the minute. Over-popularity could lead to reduced fertility in the stallion and overcrowding of the available grazing. Pressure on stallion and stud staff may result in inefficient and cursory trying with coverings attempted at less than the optimum time. Your mare might well be better cared for at an establishment taking fewer mares where more time and care can be taken. This particularly applies with maiden mares, who need knowledgeable, sympathetic handling at covering time if they are not to be badly frightened.

In nearly all cases, patience and understanding can sort out problems without recourse to any of the forcible restraints such as twitching or hobbling. On a stud taking only a limited number of visiting mares, should your mare not be quite ready for covering, or very nervous when she is first introduced to the stallion, the chances are that she will be given a long time with the horse, with the safety of the trying board between them, which will help to overcome her reticence. Then she will probably be put in a box near the stallion for the day and covered without fuss when she is fully ready. On a big, heavily patronised stud, she may well be put back in the field after a cursory trial, until the next morning, by which time the best time for service could have passed, necessitating the use of forcible restraint. I am not saying that this happens on all big studs; many are most excellent establishments, but it can occur, particularly if there are many mares coming for trial and service on the same day.

Visit the Stud

Please do make a point of visiting the stud you select, even if you have seen the stallion at a parade and picked him out as a possible sire. These home visits give you a chance to meet the stallion personally. If he chews your ears off or threatens to kick you into the next county, smile sweetly, bid the owner 'good afternoon' and do not call again.

More than this though, you have an opportunity to notice the state of the grazing and the standard of the fencing and stable management, and above all else to observe the attitude of the stud owners towards their own stock, and of the stock towards strangers. Notice if the stallion handler is wary of his charge; if he is half-frightened of the stallion and controls him with great shouts of 'Whoa!', 'Get up', 'Stand well back', etc, whilst jabbing him in the mouth and brandishing a big stick, you may be assured that that equine is in charge of the human, and not vice versa; and that is a situation that again, particularly with a nervous maiden mare, may not make for safety at covering time.

I like to see the stallion housed in the yard with the rest of the horses around him, the top door of his box open, and when he is brought out for inspection to see him standing quiet, relaxed and happy without giving the handler the necessity for constant chivvying but still retaining his true stallion presence and dignity. I would never send any mare of mine to a stallion whose temperament made him dangerous and offensive to handle.

May I make one plea for the stud-owner? When you wish to see his stud, please make an appointment. Do not just turn up— you may well arrive at an inconvenient time. Even more rude are those people who make an appointment and then fail to arrive; that is unforgiveable. Stallions will have been brushed and young stock possibly brought in for inspection, all of which takes time at a season of the year when every stud is at its busiest. If you really find you cannot keep an appointment, ring and say so. If, when you are visiting a stud, there is no one to greet you on your arrival, then wait for someone to appear or go to the house or office and make yourself known. It is exceedingly rude to go poking round someone else's stable yard, nosing into boxes, etc! There may be mares with young foals, youngsters being mouthed or even a sick animal for whom disturbance by a stranger could be a disaster. So if you go snooping about on your own, you deserve to get bitten by a foal-proud mare.

The Stallion's Qualities

One other essential about the stud you choose is that it must have a stallion holding court there who is capable of siring the sort of foal you want to breed from your mare. In other words, do not expect to sire a quiet cob with a racehorse, a show pony with a Cleveland Bay or a child's first pony with an Arab. Having made that point, it is of course up to you as owner of the mare to decide what attributes the chosen sire should have. Obviously, however, there are a number of important qualities that any stallion must possess. He must be sound and have a good fertility record; his conformation must be well above average, particularly in those areas where your mare can be faulted. His temperament must be suited to siring a foal that is to become part of the family. He should have a proven performance record and above all his progeny should be pleasing and the sort of horses you are aiming to breed. Let us consider these matters in a little more detail.

Any stallion standing at public stud must be licensed, and this should ensure freedom from inheritable defects. But it is always worth talking to owners of young stock by that particular animal and seeing if they have had any problems. If there is a recurring theme—foals that fail to thrive is an example that comes to mind—it is likely that the trouble is being passed from the sire, and that particular stallion is best passed over. I personally always avoid a stallion with bad front legs, even if tendon or other problems are said to be the result of a hard racing career. It is my experience that stallions with dubious legs tend to sire progeny with weary underpinnings too. Enquire what the stallion's fertility rate is, and how many mares he covered in the season; after all, 60 per cent fertility on 100 mares is very different from 50 per cent on two! Stud fees and keep are expensive and if there is only a 50-50 chance of getting a foal it is better to go elsewhere.

Looking at the Stallion

When the stallion is brought out for your inspection do not be so overawed by his presence (if he has not got any, he should not be a stallion) that you forget to study him as a horse. Many glaring conformation faults get overlooked in this way, so look long and carefully; note particularly his feet and legs, his set of head to neck, and of neck to shoulder, length of rein and shape of neck, length of back, depth of ribcage and power of quarters. In my experience there are two main faults which if present in the stallion are always passed on. First, bad-shaped feet, with a tendency either to being flatfooted or to being too upright and 'boxy'. The old saying of 'No foot, no hoss' is very true, and I have even heard it said that you should look at any horse for the first time with your hat pulled down over your eyes so that you cannot see above his knees. If you do not like what you see

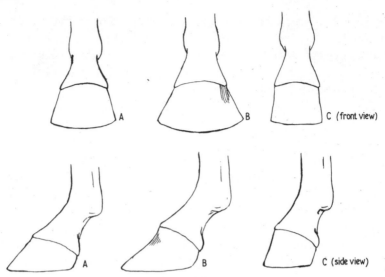

Front and side views of (A) a good foot; (B) a flat 'dinner-plate' foot; (C) boxy 'donkey' foot

there you do not bother to look further. I personally do not go quite that far but I would never wish to own an animal with misshapen hooves nor would I consider one as breeding stock.

Abomination number two is a poor set of head to neck, particularly if it goes, as it nearly always does, with a poor-shaped neck. The head should be set on to the neck by a junction, that is a graceful arch allowing the windpipe plenty of room and the animal to flex its head and neck correctly when bitted. Beware a poor join composed entirely of straight lines, particularly if it is coupled, as it so often is, with a triangular-shaped neck. Such conformation means that wind problems are likely, and a light and sensitive mouth and correct head-carriage are not.

Your chosen sire should have a good shoulder and withers, a well-sprung ribcage, deep girth and short back, leading into generous and powerful hindquarters. His tail should be set well and carried in a happy fashion. I do dislike mean, sloping quarters with a low-set tail carried clamped into the body. Good quarters should be completed by strong hocks and powerful second thighs.

(A) shows an elegant head, well set on; notice the curving join of head to neck. (B) shows a coarse head badly set on a 'bull' neck; the eye is small and 'piggy'

Ask if the stallion can be run out for you and note his movement and action carefully. His walk and trot should be long, striding, easy paces, giving an impression of much distance covered for a minimum expenditure of effort. He should track up when walking, that is, the imprint of his hind feet should fall in front of the mark of his fore feet, and he should move straight without turning his feet either in or out, in front or behind. Having made that point it must be said that an absolutely straight mover is rare; a minor degree of toes turned out in the front or moving slightly close behind is acceptable if your mare's action is all right and the rest of the stallion's conformation is good. Moving wide behind and turning the front feet in are bad faults, which can lead to dangerous, self-inflicted leg injuries, eg brushing, over-reaching and similar problems, and should be avoided.

Overall the stallion should give the impression of a proud elegant animal of well above average conformation with no glaring fault in either shape or action. His temperament too should be sensible and amenable. It is not necessary for an entire to be bad-tempered, to bite, to kick or to be unapproachable.

Having seen your stallion led out in hand, ask if you can see him ridden. If this request is greeted with disparaging remarks about stallions never being ridden, etc, again think twice about using him as a sire. It is my firm belief that all stallions can and should be ridden regularly. For one thing, a horse that is ridden cannot have too beastly a temperament; secondly, the exercise benefits their fertility and relieves much of the boredom that spoils the temperaments of so many entires who never leave their boxes except for 20 minutes lungeing a day and to cover mares.

If a stallion is ridden it gives him a chance to prove himself under saddle. After all, if you are trying to breed a riding horse, should not the sire be a proven riding horse? Both of our own stallions are ridden regularly; they hunt, do long-distance rides, go eventing, compete in hunter trials, race, are adorned with fancy dress and take part in pageants. They also go to the village for odd items of shopping. They are both capable of being ridden in any company and behave in a sensible fashion. In other words both are good all-round riding horses, and I am happy to say that their progeny are too.

I am slightly suspicious of a long string of in-hand show wins. Granted, a massive display-case of championship rosettes and trophies is very glamorous. But I am not very sure what it proves, other than that the particular horse was in the opinion of one human judge the best animal forward on that day. It does not tell you if the animal is capable of being ridden, whether

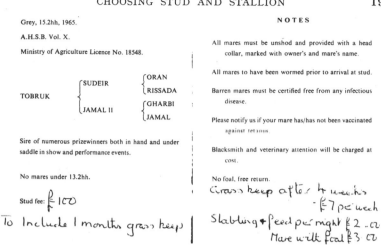

Grey, 15.2hh, 1965.

A.H.S.B. Vol. X.

Ministry of Agriculture Licence No. 18548.

TOBRUK
- SUDEIR
  - ORAN
  - RISSADA
- JAMAL II
  - GHARBI
  - JAMAL

Sire of numerous prizewinners both in hand and under saddle in show and performance events.

No mares under 13.2hh.

Stud fee: £100

To Include 1 months grass keep

NOTES

All mares must be unshod and provided with a head collar, marked with owner's and mare's name.

All mares to have been wormed prior to arrival at stud.

Barren mares must be certified free from any infectious disease.

Please notify us if your mare has/has not been vaccinated against tetanus.

Blacksmith and veterinary attention will be charged at cost.

No foal, free return.

Grass keep after 4 weeks
- £7 per week

Stabling + Feed per night £2.00
Mare with foal £3.00

Typical stud card, with additional information written in by stud owner

he will stay sound when subjected to normal work, and there is no guarantee that his good looks will be passed to his progeny.

Before leaving the stud ask if there are any of the stallion's progeny you could see. Notice if there are any conformation faults that occur regularly; if so, think twice about the stallion in question. Ask if any have come to riding age, enquire how they have done under saddle, and if they have proved quiet to break and easy to handle. Note too if they are confident enough to approach a stranger—timid, nervy youngsters often make unpredictable adults.

The Financial Arrangements

Before you leave the stud ask for the stallion's card. This should tell you his stud fee and what conditions attach to it; it will also tell you what happens if the mare is barren. The most usual terms are 'No Foal Free Return' (N.F.F.R.); this means that should the mare not get in foal, she has a free service in the following year. Please realise that normal grazing and keep charges will still be payable. 'No Foal No Fee' (N.F.N.F.) means that should she be tested barren by 1st October, no stud fee is payable. 'Fee £75 straight' means that the stud fee is payable whether or not the mare is in foal. The Stud card should also give the weekly rates for keep at grass, extra feeds, stabling at night, and what swabbing is required. If any of the information is not on the card, ask—it saves any embarrassment later. If perhaps the charge for grass keep is not written down, ask the stud owner to make a note of it for you on the stud card, likewise anything else you wish to know. I say this, having in the past had mare owners complain when they have been charged for perhaps two or three nights' stabling for a mare and foal during a period of bad weather, after agreeing that their animals should be brought in out of the wet if necessary.

Standing a stallion at stud is not a licence to print money. It involves much hard work, patience and very long hours. The job is not without its risks when handling difficult maiden mares or foal-proud matrons. It is not possible for the stud owner to act as a charity institution for mare owners, and so you must expect to pay for all services rendered over and above the normal covering. In other words your stud bill could include as well as the stud fee, grass keep, extra feeds, stabling at night, veterinary attention, blacksmith's charges, worm powders and the stud groom's fee, usually between £2 and £5. Do not forget either that your bill may be subject to Value Added Tax.

The covering certificate is normally forwarded with the receipted stud account, but if it does not arrive ask for it and then preserve it carefully until the foal is born and registration is due. As a matter of interest, we get asked for duplicates of at least 50 per cent of our covering certificates each year. 'I put it away so carefully that I can't find it' is the most usual excuse!

Sending Your Mare to Stud

Having gone home and thought carefully about the various stallions and establishments you have seen, make your decision, and having made your decision stick to it. I have known mare owners book into four different establishments because they didn't know which stallion to choose! Make your booking, get any necessary veterinary checks done and send your mare off to stud about a week before she is due in season. If you don't know when that is don't worry, the stallion will, so send her and let them decide. I mentioned swabbing. Your stud will advise you on what veterinary tests are required to ensure that your mare is free from infection. Some studs prefer to get their own vet to carry out this testing, but if it is necessary for your vet to do it, then ask him in good time so that you are not delayed in sending the mare to stud.

A word here against covering mares too early in the season. Nature did not intend foals to be born in January or February, nor did she intend mares to come in season then. The time for young things to arrive is when the spring sunshine is beginning to be warm and the new grass to grow, in other words from about the middle of April. This means that your mare does not need to go to stud before the beginning of May at the earliest.

Your mare should arrive at the stud, wormed within the previous week, with anti-tetanus injections up to date, roughed off if she has been in all the winter, all her shoes removed and her feet pared and wearing a supple leather headcollar. Please warn the stud of any idiosyncracies that the mare may have, such as being difficult to catch or terrified of sheep. Deliver your mare to the stud after making arrangements as to time and date (do let the stud know if you are going to be late or can't make it on the appointed day), unbox the mare, give her to the stud staff and *go away and wait to be told that your mare has been covered. Do not pester.* Remember the stud know their job and it is in their interest to get mares covered, in foal and home again, as quickly as possible.

If you hear nothing after three weeks it is quite permissible to ring and enquire what is happening. If your mare is proving difficult to get in foal or into season and the stud suggest veterinary treatment, do be guided by what they say. They almost certainly know more about the matter in hand than you do.

What Will a Stallion Pass on
When Used for Cross–Breeding?

What attributes will stallions of the various breeds of horses and ponies pass to their progeny if they are used for cross-breeding? Pure-breeding is not considered here, because I have already listed the points that every good sire, regardless of breed, should have. Breed type and characteristics are better assimilated from society publications and from watching as many show classes for the breed of your choice as you can. The native ponies such as the Dartmoor, Exmoor and Welsh, have many sterling qualities which are well known to their devotees, but the stallions of these breeds are too small to be of much use in a cross-breeding programme, with the possible exception of the Welshman on whose bloodlines the English show pony has been founded with judicious admixtures of Arab and thoroughbred blood.

The mares of the small native breeds, however, if crossed with an Arab and the female progeny retained and top-crossed with either an Anglo or part-bred Arab or thoroughbred, can be expected to produce, in the third generation, a versatile riding horse which is likely to combine the quality and speed of the thoroughbred, the presence, soundness and stamina of the Arab, and the substance, sense and cleverness (some call it cunning) that is possessed by all native ponies. We maintain that this admixture of native pony blood provides the necessary cross of commonsense.

I should just mention, however, that in my opinion Shetland ponies should never be crossed with any other breed. As a pure breed they are superb miniature draught animals, the driving pony par excellence, but their individuality and 'carty' conformation do not make for an attractive cross in either make or temperament.

The stallions of the New Forest, Connemara, Welsh Cob, Fell and Dales breeds are likewise not much used in cross-breeding, although if one owned a pony mare lacking in substance one of these breeds might just be considered as a suitable sire. The trouble is that, used as a top cross, often the substance appears in the form of coarseness, plain heads and thick necks. The females of these breeds, crossed in the same way as the smaller native breeds will, however, produce the same quality riding animal in the third generation, albeit in a larger edition—no bad thing in these days of ever-lengthening children!

The next major sector of the breeding scene is the show pony and show/working hunter pony. What advice am I to give? Elegant little beasts float round show rings for the whole of the summer with ever more competent children appearing on the scene to pilot their progress. Frankly they terrify me and I confess that they all look alike! I think the only help I can give is this: there is such enormous competition at all levels in the show-pony world that unless you have a pony mare that has won regularly in the ring, or is closely related to one that has, you are possibly better advised to aim to breed something with a bit more substance and sense, that can earn its keep in the hunting field or as a pony club and general fun pony. The British show pony has been most carefully bred over the last forty years and is rapidly emerging as a true-breeding new breed of pony, so if you are seriously intent on joining the exalted ranks of successful show-pony breeders, then seek the help of a reputable breeder whose ponies do their share of winning, and learn from them what is required.

The Arab is one of the breeds most commonly used for cross-breeding. Everyone has a mental picture of the gorgeous mythical beast of the desert galloping across the sand, head and tail in the air, with a handsome white-robed sheik on his back, but not so many people really know what qualities the Arab has to offer. Over the years much prejudice has arisen both for and against the breed. Some Arab enthusiasts have only themselves to blame for a great deal of the anti-Arab mutterings that can be heard in hunting and eventing circles.

Let there be no mistake: the Arab is the most potent crossing sire there is, capable of stamping his courage, presence, soundness and stamina upon his offspring even when they are out of the most unlikely dams. He himself is a tough working animal capable of carrying considerable weights for very long distances over difficult terrain under extreme climatic conditions. The trouble is that he is also a very beautiful animal, and because of this many breeders have kept Arabs for show purposes alone. Indeed we have the situation on some studs where none of the animals are broken to ride or ridden, and stallions standing at stud are bred from parents and grandparents and even great-

grandparents who have never proved their abilities under saddle
The result is that faults of conformation and temperament have
crept in and become established because they have not been
shown up as weaknesses when the animal has been subjected to
the normal stresses and strains of a career under saddle. For
instance, a nervous, flighty temperament is of no consequence if
the animal never leaves the stud paddocks, but if someone buys
a gelding from this blood line and tries to ride it, accidents and
troubles result. The unfortunate owner then goes rushing back
to the reliable if less-inspiring hunter type with a cry of 'Ugh,
Arabs! Never again!'

But if you select as a sire an Arab who is ridden regularly and
whose stock and antecedents have proved themselves under
saddle, you should find that his part-bred progeny grow up into
just about the best general-purpose riding horses one could wish
for, capable of an immense amount of work once they are fully
mature: a kind and willing ride, sensible and sure-footed, with
good paces and action, presence and courage unlimited, and
generally possessed of the sort of temperament that makes them
nice people to have around.

The only thing that those who have no experience of riding
the Arab and its crosses may find is that they are Arab-quick.
This is to say that their paces are quick and active and that,
combined with the fact that they are all more intelligent than
other breeds, does mean that anything they decide to do, be it
sinful or otherwise, happens in about half the time it takes any
other breed to think about it. This is no disadvantage if you've
been warned and have learned to adapt: in fact anything
without the Arabian sparkle comes to seem dull.

It must be said, though, that not everyone does learn to
adapt. I once bought a very nice part-bred Arab mare cheaply
because her owner said that she ran away with her at a walk!
The little mare had indeed a very fast active walk, and when
allowed to stride on could easily walk at 6mph. The trouble was
that her ex-owner had been used to a placid cob that she had to
kick along. Sitting on an on-going mare she had started to hang
on to the reins to steady the mare back to the pace of her old
cob; the mare had resented this, not unnaturally, and started to

throw her head about, and that had resulted in the application of more severe bits, martingales, etc. Oh dear! I had better say that Arabs and their crosses tend not to make horses for the faint-hearted.

This is particularly true of the Anglo-Arab, which is the straight cross between the Arab and the thoroughbred with no other blood in the pedigree. The Anglo-Arab is the performance horse supreme, having immense natural athletic ability combined with the speed of the thoroughbred and the stamina and presence of the Arab. A warmish mixture this that thrives on hard work, but unlikely to make suitable mounts for novice or weekend riders. So, if you have a thoroughbred mare, I would suggest using either a hunter sire or a big part-bred Arab, rather than a thoroughbred or Arab stallion, if you want to breed the sort of horse who will happily live in the field all the week and then be busy over the weekend without complaint or getting overfull of himself.

A big part-bred Arab is likely to sire a very useful type of riding horse from either a thoroughbred mare or a hunter-type mare with a lot of thoroughbred blood in her makeup. This cross will not have such pronounced Arabian characteristics, and indeed if you are breeding a part-Arab you do not want to end up with an animal resembling a pure Arabian.

To be eligible for registration in the part-Arab register, an animal's pedigree must contain a minimum of 25 per cent of Arab blood, so if your mare has no Arabian blood in her own pedigree, the stallion of your choice must have at least 50 per cent Arabian blood. One other thing: if size in your hoped-for foal is important to you, investigate the stallion's pedigree and check that he is in fact bred from horse stock, even if he is 15.2 hands plus himself. There are many instances of big stallions throwing back to pony ancestors. The ultimate size of the stallion's progeny will also prove a useful guide as to the height your foal might attain.

Many of the Eastern European riding-horse types, such as the Trakener, Wielkopolska and Malapolska, are in fact big part-bred Arabs that have been bred within the limits of the original stock and its derivatives so that over the course of time they have become self-perpetuating breeds capable of stamping their type on their progeny. Many of the top continental show jumpers, dressage horses and eventers are in fact bred this way.

There is one important fact that has helped these breeds to establish themselves. All the breeding stock is performance-tested. Potential brood mares and stallions have to pass rigorous selection tests, attaining set standards in dressage, jumping, speed, endurance and not infrequently harness work. All must be absolutely sound in both mind and body. A superb animal whose temperament made him unsuited for ridden work would be culled, and not admitted to the breeding programme.

We come now to the thoroughbred, possibly the most widely used sire of all and certainly the best-known. The thoroughbred is a very high-couraged animal bred to race and jump at great speed, but regretfully in the selection for speed, faults of both temperament and conformation have been overlooked, and so one finds that problems caused by poor limbs, weak backs, poor feet and respiratory deficiencies are more prevalent than they should be. There is also the fact that the more highly strung animal is likely to make the best racehorse, and so by constant selection for speed and racing ability the average thoroughbred has come to have a temperament that tends towards the neurotic.

However if you are aiming to breed a big hunter or event horse you will probably have to opt for the thoroughbred sire in order to get the size and scope.

The problems that beset the modern thoroughbred are not helped by the fact that many of them are broken and ridden as yearlings, are on the racecourse at two years old, and some colts are retired to stud at three. This use of 'baby' horses throws great strains and stresses on immature limbs and bodies, and so one has no way of knowing if the weaknesses thrown up by a racing career were caused by that racing career, or if they would still have occurred had the animal undertaken a more normal life as a riding horse. So, when choosing a thoroughbred sire, choose if possible an older horse who already has progeny who are performing well under saddle and who, above all, are staying sound and non-neurotic whilst they do so. It is always worth travelling a bit further and paying a little more in order to use a proven stallion.

Thoroughbreds and their crosses tend to be the most expensive of all to rear to maturity and if your youngster is by a proven sire it is more likely to recoup some of the money expended upon it. Thoroughbred youngsters grow bigger sooner, eat more, are more accident-prone, have thinner skins and therefore cannot really live out at all in winter, and above all you need to think for them, as their inbred instinct to gallop stops short at telling them when it is safe to gallop, and when a slower pace is more prudent. On the other hand though, a gallop on a really fit, schooled thoroughbred is an experience to be savoured; the effortless stride, the feeling of power beneath you ready to be released in dynamic action at a mere relaxing of the little fingers, the feeling of oneness between yourself, the horse and the atmosphere through which you are travelling at anything up to 35 miles an hour, bring you close to the sensation of flying on your own wings.

Most hunter mares are eligible for Weatherby's non-thoroughbred register. The conditions for entry are somewhat complicated, but it is always worth enquiring from the registration authorities if, firstly, your mare is registered—many are, and the fact goes unrecorded when the animal changes hands; or secondly, if she is eligible for registration. Once the mare is accepted into the register her foals by a thoroughbred stallion will also be eligible for registration, and there is no doubt that a registered animal is that much more valuable because the date of birth and pedigree particulars can always be checked.

Whilst on the subject of registration, when you eventually apply to have your foal registered, you will need a covering certificate to show that your mare was indeed covered by the stallion you have named on the registration application form as the sire of the foal. This must be signed by the stallion owner.

Some of the light carriage breeds, such as the Cleveland Bay and the Hanoverian, are sometimes recommended as sires of riding horses. There is no doubt that they throw progeny with great size and substance, but they are of cold-blooded (cart-horse) origin and it is occasionally said that some of their crosses

| CERTIFICATE OF COVERING OF MARE | Nº 2595 /79 |

MARE		STALLION	
Name			
Name of Owner at time of covering		Name of Stud at which standing	
Address		Address	
USE Block Capitals			
Name and Address of New Owner if Mare sold after cover-ing		Signature of Stallion Owner or Representative	
		DATE OF LAST SERVICE	
		Signature of Mare Owner or Representative	

Covering certificate issued for a registered mare

lack speed and courage when faced with the demands of a long day's hunting or big cross-country fences.

It may be gathered that my own preference is for the Arab and his crosses, in particular the big part-bred riding horse. But after I've said my piece, and everyone else has extolled the virtues of their favoured breed, the final choice is still with you, the mare owner. Get a clear picture in your own mind of what sort of horse you are trying to breed and look at as many different stallions and their progeny as you possibly can. Look at the dams, too, and try to learn as far as you can (heredity is an unpredictable beast) what sort of mare produces what when mated with stallions of varying breeds.

Brood-mare classes at shows are often most revealing on this aspect of breeding. Remember too the performance angle: a horse that is not sound in either mind or body is no profit to anyone but the butcher.

Make up your mind and stick to your own decision; don't be swayed by either friends or fashion. Remember the old breeders' adage: to the best from the best. And hope for the best.

3
The In-Foal Mare and Preparing for Foaling

You will be advised by the stud that your mare has gone over her three weeks and is ready for collection. This means that having been served, she has not come in season again three weeks later and may well be in foal!

Either leave her for six weeks, if you want to be as sure as possible that she is in foal, or else go as soon as you can to collect her. I say this because it is not a good idea to travel a mare or upset her in any way (this includes the excitement of showing) between the 30th and 40th days after service. The reason for this is that somewhere about the 33rd–35th day after covering, the embryo (fertilized egg) which has been floating loose in the mare's uterus becomes implanted in the uterine wall and the true placenta starts to develop. The development of the embryo before implantation is maintained by hormones within the uterine cavity, but at implantation this maintenance is taken over and continued by the developing placenta. So for two or three days the entire future of the embryo is very finely balanced as the uterus adjusts to its gravid state and the placenta begins to secrete the hormones which will ensure the continuance of the pregnancy.

It is obvious that any upset at this time—travelling, meeting strange horses, showing, even a change of diet—can upset this equilibrium and cause the fertilized egg to be lost. So keep your mare very quiet with no stress and no change of routine until 40 days have elapsed since she was covered.

Having got her home, she should be turned away in a field of good grass for at least the first eight weeks of her pregnancy. For preference, she should have the company of another mare or mares. Another in-foal mare is of course ideal, providing they agree. An oft-repeated horsy old-wives' tale is that an in-foal mare should not run out with a gelding in case the gelding forces his attentions on her and causes her to lose her foal. I remain unconvinced that there is any truth in this. Many more foals are lost by poor nutrition in the mare in the autumn, and also a great many of the mares said to have been caused to abort by the attentions of a gelding were probably never in foal anyway. None the less, female company is to be preferred, particularly for the first eight weeks after covering, and it obviously goes without saying that any uncut colt or gelding with riggish tendencies should be kept well away from in-foal mares at any time. After the eight weeks have passed your mare may be quietly ridden if you wish, but there must be no jumping, no long-distance rides, no eventing, no hunting and not too many shows. These activities must stop by the beginning of November or even earlier if the expectant mum seems unwilling to work.

The Mare's Diet and Housing

The mare's nutrition during this period is important. It must be adequate to keep her in good condition and to support the growth of the foetus, although she should not become grossly fat. Be particularly careful as autumn approaches. Many foetuses are reabsorbed at this time due to a rapid fall in the mare's nutritional level as the goodness goes out of the grass. This occurs at the same time as the foetus is enlarging very rapidly. Reabsorption is nature's way of protecting the mare from being harmed by the demands of a developing foetus if her bodily reserves are insufficient to meet them. You will not be aware that it has happened until the mare suddenly comes in season about three or four months before she was due to foal!

So it is a good thing about August time to bring your mare in out of the worst of the flies and give her a feed, which should be increased gradually to two corn feeds a day by the beginning of October. Make all feeding changes gradually; a pregnant mare may at times be fussy to feed, so humour her along and increase amounts gradually.

Here I will mention the very prickly subject of mineral supplements. We do not use them; the only addition we feed is ordinary salt, either as rock salt in the manger or added to the feed—say a dessertspoonful once a day. We also firmly believe that mineral additives fed in poorly monitored doses by unskilled hands can cause all sorts of imbalance problems. A balanced, varied diet with the addition of an occasional linseed mash, sliced carrots, sliced apples, black treacle or molasses, etc, should supply everything that is needed. If you really think you have a problem ask your vet, who will be only too happy to advise. Do not listen too closely to the feed firm's representative, who may get paid by commission on sales.

Your mare should be in at night by Christmas at the latest, and having two good feeds a day and her usual ration of good hay. Do beware of feeding dusty or mouldy hay to a heavily in-foal mare, as not only is it lacking in feed value but the dust and inevitable fungus spores can cause abortion and future infertility as well as respiratory problems.

Make sure as well that she has a good deep bed; pregnant mares like to lie down (or should) quite a lot, and the depth of bed is needed to support their extra weight and to prevent slipping when they rise.

Turn your mare out every day for exercise unless it is particularly cold or wet, or very slippery and icy. Just an hour or so is enough if the weather is poor. Remember to keep the mare's feet trimmed; she has enough extra weight to carry without having to hobble about on long, ill-shaped feet. And it is obviously more important than ever to attend to her regular worming programme. Do, however, make certain that whatever product you use is cleared for use in pregnant animals; ask your vet if in doubt. She should also be given an anti-tetanus booster about six weeks before foaling so that maximum protection can be passed to the foal in the colostrum.

How the Foetus Grows

To know what is going on inside your mare, even if only vaguely, is useful. It will give you an idea of how much nourishment the mare will need to supply to the growing foetus and why she may look rather hollow after foaling.

The cell that will develop into a foal is no bigger than a pinhead at conception. Only one sperm joins with the egg at the moment of fertilisation. After fertilisation the outer cell wall of the egg becomes impervious to further sperm. The egg cell first divides into two, then four, then sixteen cells, and gradually specialist groups of cells appear which will eventually become organised into the various parts of the embryonic horse.

At 40 days after conception the embryo has lost its spherical shape. Limb buds and the head can be seen and the eye-spot is present. The embryo or foetus is now about 5cm (2in) long. The embryo and its membranes are about the size of a tennis ball at this stage. The true placenta has formed and attached itself to the uterine wall and will grow apace with the foetus, to provide it with the nourishment necessary for growth and development.

By 80 days the foetus has trebled in size. The limbs are more or less complete but immature. Hooves are discernible.

At 120 days after the fertilized egg started to divide, all the physical features of the foal are present—ears, eyes, nostrils, limbs, hooves, a rudimentary tail and all internal organs. In short, the foetal foal is a 'complete horse' but very small, immature, and quite incapable of independent life. At this stage it is still naked.

After this stage, the foetus is safe from deformations of the thalidomide type. We once had a superb filly foal born alive but with no front legs, only rudimentary flippers. The dam, whom we had purchased in foal about six months previously, had—we discovered after many intensive enquiries—been caught in the spray drift of 2,4,5,T. brushwood killer some forty days after service, just when the limb buds were forming. So perhaps I should add one further piece of advice: keep your in-foal mare well away from any drifting sprays being spread over fields or gardens, no matter how safe the makers claim they are, preferably at all times and certainly until the first 120 days of pregnancy are well past.

After the 120-day stage, growth is quite rapid. The foetus is 40cm (15in) long at 150 days, 55cm (22in) at 210 days and 90cm (33in) at 280 days. Growth becomes most rapid in the final weeks of pregnancy. Not only does the foetus increase in length, it puts on weight. (A thoroughbred foal weighs 100-120lb at birth, Arabs and their crosses rather less.) Its internal organs mature and it grows a coat, mane an tail. In other words, nature is making final preparations for your foal to be born into the world alive, in a condition enabling it to be capable of staying alive and thriving. It follows from this that a foal born prematurely is rather more at risk than a full-time foal, and obviously the risk becomes greater the more prematurely the foal is born. It is usually considered that a foal born six weeks early has little or no chance of survival.

It is equally true that a foal that is carried for more than twelve months is at increasing risk. Remember, though, that many thoroughbred mares habitually carry their foals for twelve months instead of the usual eleven: in these cases no harm will be done to the foal. It usually seems to be after a long, severe winter that you hear of foals being carried for as long as fourteen months and then being born in an overgrown but emaciated state. These foals rarely survive and indeed their actual birth can present problems. The cause of such overlong pregnancies is not known, but it may be the result of a poor level of nutrition in the mare, lack of sunshine or lack of Vitamin D: it seems as if the normal development of the foetus is delayed whilst its bones go on growing.

During its period of development in the uterus, the foetus obtains its nourishment through the placenta. This remarkable organ develops from the embryonic membranes and adheres to the wall of the uterus in such a way that the gasses and food-stuffs necessary for the growth and development of the foetus, eg oxygen, carbohydrates, proteins, can diffuse from the mother's bloodstream into that of the foal. Also, waste products and gasses, such as carbon dioxide, can diffuse from the foal's bloodstream back into the mare's and be eliminated through her lungs and kidneys.

Foal the Mare Indoors

Some six weeks before the mare is due to foal she should be moved into the box where the birth is to take place, if it is not her usual one. This gives her system time to make anti-bodies against the specific bacterial flora of the box. These anti-bodies are passed on to the foal via the colostrum at birth and are a much better protection against joint-ill, scour, etc, than all the veterinary treatment in the world. This is where the one-mare owner scores. On big public studs, mares are never in the foaling boxes for more than 24 hours and problems do occur.

This brings me to another point; for this reason alone, even if you are inexperienced, it is always worth foaling your mare at home, amongst her own bugs so to speak, rather than rush her off to stud to foal.

At this time it is probably worth putting your mare on to four small feeds a day rather than two big ones. The foal increases in size very rapidly towards the end of the pregnancy and the dam may well be too full of foal to eat a big feed and not clean up. Four little feeds, the biggest of the four being given as late in the evening as possible when all is quiet, will encourage her and help her to assimilate the maximum goodness. You may well find that she cuts down on her hay consumption too.

Obviously clean water must always be left with her. Pregnancy puts quite a strain on the kidneys, as the mare is responsible for excreting two lots of waste products through her own system; it is essential that she has plenty to drink to provide a good flow of fluids through the kidneys, keeping them clear of any accumulation of toxins.

About a fortnight before the foal is born it will move into the birth position. At this time your mare may well show signs of unease, go off her food and to the inexperienced give every sign of imminent colic; this phase lasts often for two or three days, after which she will return to normal. The only danger is that one may mistake a pre-foaling discomfort for colic, or worse, a case of genuine colic for the other; I have known vets who have treated perfectly fit mares with drastic remedies for colic when all that was happening was the foal turning. Remember that genuine colic has a usually reliable symptom: the sufferer will not touch either food or water. But in this as in all things, if really worried consult your vet—who should of course, and I cannot repeat this too often, be a *horse* vet.

You may wonder why I am so adamant that your mare should give birth to her foal in a suitable box rather than out in the field, as one often sees advocated. There are several reasons. Firstly, the mare can be kept under observation in comparative comfort. No one is going to traipse out into a dark, cold and possibly wet field at fifteen-minute intervals to see how the expectant mum is getting on. You will not go out to her as often as you should, or not go at all, and people blame anyone but themselves if the mare foals in an awkward spot and injures or kills the foal in the process.

Another most important reason for having the mare in a stable is that should there be any abnormality at the moment of birth, the vet is in problems if he is struggling by torchlight in long wet grass and streaming rain. If the mare is in a stable with electric light and a clean, dry bed, he has a much greater chance of being able to put matters to rights. Mares in the process of giving birth are naturally solitary animals and even if in difficulty they can prove impossible to catch; a mare can go rushing off around the field with a half-born foal stuck partway through her pelvic girdle with, needless to say, considerable risk to both herself and the foal.

Suppose, too, that your mare foals outside on a cold or wet night. Just imagine the shock the new-born foal receives coming from inside the warm uterus into a temperature that might be in the low forties. There must be a risk of the foal failing to maintain its body temperature and dying of cold and shock.

There are other risks to a new-born foal in the open. Carrion crows have been known to peck the eyes out of partly born young animals, foals have been trampled to death by other animals coming to see what is going on and they have also fallen into ditches, rolled under fences, lost their dams, etc. The horrendous stories I have heard are legion. All were avoidable had the mare foaled inside. Think too of the maiden mare with a ticklish udder who refuses to let the foal feed, or the mare who decides she doesn't like her foal: such situations are more easily brought under control in a loose-box.

There is one other vitally important advantage that to my mind almost outweighs all the other things: if the foal is born indoors with you, his owner, present and supervising the birth, the first thing he will see apart from his dam is you. From that moment human beings, and you in particular, are part of his world, implanted in his baby brain, and he will forever be quieter and easier to handle.

We have proved this implantation theory beyond all doubt. The receptive time, when things become most impressed upon a young animal's mind, is for about the first three hours after birth. Just occasionally one of our mares has foaled unexpectedly in the night with no one there, no human seeing the foal until the following morning. These foals are never as quiet or trusting to handle, and a foal born out in the field whose mother cautiously takes it away when the owner approaches is even less likely to be favourably impressed!

It is a good idea to give the box chosen for the mare to foal in a thorough springclean before she moves into it. All cobwebs and other dirt and dust should be hosed away and the floor thoroughly scrubbed, with Jeyes if you like; after everything has dried, the walls should be given a clean coat of whitewash. Check too that the light works, the switch is safe and properly guarded, and that the light bulb is properly guarded and cannot cause a fire hazard when the light is left on all night. If the manger is a square metal one fixed to the wall it must be removed, so that the foal cannot injure itself on it in its first struggles to get to its feet. Make sure too that there is not a gap under the door that is large enough for a foal to get its leg caught. Check that the top door is usable—I have known them fall to bits when moved—and that it will shut and fasten. It goes without saying that the door of the box must open outwards, but then all doors of buildings housing livestock should do this.

Once your mare is installed in the foaling box leave the light on at night. It only costs a few pence, you can then see in a minute what is happening without disturbing your mare with sudden illuminations, and should she catch you out and foal unexpectedly, both she and the new-born baby can see where and what the other is. It may sound silly but I have known maiden mares to be absolutely terrified of their new babies, and a little light on the scene may help to ease the situation.

For the same reason, we leave a headcollar on the mare once she is within about a month of her due date. It is bad enough trying to cope with a panicky, frightened or possessive mare without having to catch her to get a headcollar on her first. Let it be a well-fitting supple leather one, though.

Remember the Paddock

The paddock that the new-born foal and its dam are to use should be rested from all stock, if possible, for at least six weeks before the foal is due. Six months is better, but is not usually practicable on small establishments. A little pasture management will help to make the six weeks' rest much more effective: all possible droppings should be picked up and removed, the ground should be thoroughly harrowed and then rolled and a light dressing of a seaweed-based fertilizer given. The long coarse patches should have been topped off in the autumn and it is helpful if some bullocks can be imported to graze over the paddock at this time. Mixed stocking, with cattle following the horses, is one of the best forms of worm control.

A friend who has a small stud has a very ingenious method of picking up droppings and topping off uneaten coarse grass and weed: she asks a neighbouring farmer to come in with his forage harvester and dung spreader. The forage harvester picks up the droppings and cuts and pulverises the weeds. The droppings and pulverised weeds end up in the dung spreader, and the farmer then spreads them on his corn stubble, where they act as manure for the subsequent crop. This depends on having a co-operative farmer friend, but it does show how a reasonable standard of pasture management can be attained.

Fences should be attended to, and any potential foal-disaster-spots fenced off. One word here about ragwort: this horrible weed should never be tolerated in any field where livestock graze. It should be carefully pulled, roots and all, and burnt.

4
Foaling

Once your mare shows signs that the birth of her foal is imminent, in other words once she has made an udder and the foal has positioned itself for birth, it is essential that she is kept under close observation at all times. She must still get her regular exercise, but two or three hours in the paddock daily will suffice. It is rare for a mare to foal during daylight hours, but there are no hard-and-fast rules.

Her box should routinely be bedded up for foaling each night, then if she does foal without warning no harm is done. It is a good idea to keep the box floor sprinkled with a little sand under the bedding; birth liquids are very slippery—they exist to lubricate the passage of the foal to the exterior—and a light scattering of sand (two or three handfuls are more than enough) helps to keep the bedding straw in place and prevent the mare slipping should the floor become exposed during the birth. The box must be well bedded with clean straw, which should go right up to the walls and be banked against them and should also reach to the door and be sufficiently deep to prevent draughts penetrating under the door.

The light should be left on during the hours of darkness. Supervision of your mare is then far easier because you can observe what is happening without having to enter the box. Indeed if the chosen box is conveniently positioned it may even be possible to keep watch from the warmth of the house.

Most mares lie down to foal and should she get to the stage of delivering the foal's front legs unobserved, it is possible for them to be broken if she is startled and jumps to her feet. The sudden switching on of the light is more than enough to make a tense and worried mare do this. There is also the risk that she might slip and break her pelvis.

Remember that a supple well-fitting headcollar should

always be left on your mare when foaling is imminent. The quietest, kindest mare may become a raging, foal-proud lunatic once she has foaled and it is not easy to catch her and put on a headcollar in those circumstances. Much better to have something ready to grab if necessary.

Once your mare is at the stage when foaling is obviously close, she should be brought into her box by about 6 pm, given a little feed and left quiet with some hay or a rack of fresh-cut long grass (*not* lawn mowings as they heat and are dangerous). Her main feed will be later on, at about 9 pm, so that you have a chance to take a last look at her before you go to bed, at about midnight. Until signs of imminent foaling are seen, there is no need to stay up with her every night. If you adopt the bring in at 6 pm—small feed—main feed at 9 pm—check again before bed routine, you have every chance of noticing any rapid changes in the mare's condition.

The signs that indicate that birth is near are, obviously, a full udder and, more reliably in our experience, a dropping away of the muscles at either side of the root of the tail.

Taking these points in more detail, the mare's udder will first start to increase significantly in size some six to eight weeks before the foal is due. This increase in size will be gradual, but much more rapid as birth approaches. Some mares will also get oedema (swelling) in front of the udder, along the lower line of the belly, so tight and tense does their bag become. Most authorities set great store by 'waxing' as a guide to when a mare will foal. Wax is the little blob of white waxlike substance that hangs from the teat like a little white icicle. It usually forms about twenty-four hours before the foal is born. The udder becomes shiny and moist looking, wax appears on the teat, then eventually drops off, and milk begins to drip away. Usually the foal is born within about twelve hours of this, but we have one mare who always runs milk for a fortnight before foaling and another who never waxes until after she has foaled.

We find that the sinking of the muscles around the tail head is a sure indication that birth is imminent. If you stand behind a horse you normally see, as the illustration shows, powerful rounded muscles that go right up to and surround the root of

(left) Normal horse from behind—muscles are tight to tail head; (right) an in-foal mare with birth imminent—tail-head muscles have slackened

the tail. When the mare is about to foal, the hormones that control the birth process cause the ligaments of the pelvic region to become lax and flexible, so that the bones of the pelvis can move far more than normal to allow for the passage of the foal. This relaxation of the ligaments means that the muscle masses of the quarters are not held tightly to the bones as normally, and so the muscles drop away from the spine and the tail-head stands proud, well above the rest of the muscles of the quarters. The hollow alongside the root of the tail and end of the spine is usually quite pronounced and obvious once you have seen it.

So if your mare has waxed up and her muscles have dropped away, you should start sitting up with her.

If your mare starts to drip milk, even if it is only twelve hours before she is expected to foal, it is a very good idea indeed—if she will allow you to handle her udder and you can milk—for you to draw off about 300-500ml of this first milk and put it, in a previously boiled and cooled glass container, into your deep freeze. This first milk, or colostrum, contains all the antibodies which will protect the foal against infection for the early weeks of its life, and if the mare runs milk for even a very few hours a great deal of the precious liquid will be lost; the foal will then lose valuable protection which is much more efficient in coping with the specific bacterial flora of your establishment than all the antibiotics that have yet been discovered.

There is also the point that should the foal not be able to reach the mare's udder for any reason, or the mare not be able to allow the foal to suck, you can give the foal this vital colostrum from a bottle, if need be. Of course we hope that this will not be necessary, but insurance can be a wise thing to have.

Having said this I will hastily add that many mares will not allow you to milk them, and obviously it is not worth risking a terrible upset; and if you have not learned to milk, now is not the moment. In any case you must have someone to hold the mare for you whilst you make the attempt; proceed very gently with firm but soft movements, talking soothing sweet nonsense all the time. Don't tickle and don't poke, and don't try and milk with long sharp finger-nails. Think about it and most people would agree that they deserve to get kicked! Finger-and-thumb milking is the only way with a mare; the teats are very small compared with a cow's. If all your gentle, well-intentioned movements are greeted with snapping teeth, swishing tail and cow kicks, leave it and take consolation that these sort of mares usually make the best mothers.

We had a beautiful Arab mare who was very fond of human beings; she would follow us like a dog and thoroughly enjoyed our company. She was duly covered and produced her foal at the appointed time. Then the fun began. We could milk her quite happily, but it took three thoroughly competent adults all their time to restrain her sufficiently to allow the foal to suckle. She did eventually allow the foal to feed without someone in attendance, but the youngster's access to the milk-bar was severely rationed and we had to wean him at four months old. The mare returned to her career under saddle, as she was obviously unsuited to being a brood mare. On the other hand our most successful matron has never allowed anyone to put as much as a finger on her udder, and she has to date reared eight foals to large and bouncing maturity.

The task of sitting up with your expectant mum is a most exhausting one and is much improved by being shared between two or more people. On our own establishment my husband and I always share the night's vigil. I am a much better night owl than he, so I do the first half of the night, from midnight to about 4 am, and he does the early watch, from 4 am to 8 am. There is no need to spend the entire time draped over the loose-box door like a semi-conscious poltergeist! Aim to visit the mare every hour. If you are short like me put a solid box or bale of straw outside and to one side of the loose-box door, so that you can stand on it and get a good view of what is happening without disturbing the occupant. There is no need to enter the box; indeed you will get a better notion of what is going on if the mare does not know you are there. Observe quietly, and if nothing is happening creep away and return to your book and the coffee pot.

If when you go out you find that your mare is not eating or dozing, be prepared for action—particularly if there is any restlessness or tail-switching. The first signs that the process of birth has started are the swishing of the tail, wandering about the box, possibly getting down and getting up again—sometimes with horrifying grunts and groans. It is not unknown for mares to reach this stage, get up, have a good shake and postpone matters until the following night. However the chances are that once matters have progressed this far you won't have long to wait to find out if your foal is a colt or filly.

The important thing at this stage is for you to do nothing at all. Stay outside the box, keep quiet, keep still and observe. Foaling is a completely natural process and will occur in 99 per cent of cases without human assistance; and even if it does not, your job as an untrained amateur is to observe that all is not well and summon professional assistance quickly and urgently.

One word of advice: this observing task can be exceedingly cold for the sitter-up. Do make sure that you are warmly dressed. You are of no help to anyone or anything if you are paralysed with the cold. Let the wrappings be old ones though; you can get very wet and very slimy-dirty if by chance you end up clutching a large, soaking wet, slippery, just-born foal, perhaps to pull him round to the mare's head.

One other warning: as I have already said, birth fluids exist as lubricants and as such they are most efficient; if you wear a ring of any kind, take it off before things begin to happen in earnest. It is possible for even the tightest-fitting ring to slip off your finger unnoticed if your hands become covered in birth liquids. I once left my wedding ring inside a cow, but that's another story!

As I say, the first signs that foaling has commenced are that the mare shows signs of discomfort, swishes her tail, stops eating, puts her ears back and may start to sweat slightly. This stage will usually continue for about half an hour and then the actual birth process will start with 'the breaking of the waters'. The placenta which has been surrounding the foetus and its membranes ruptures, allowing the chorion or water-bag to be expelled via the mare's vulva, when it breaks. Once the water-bag has broken, the process of foaling is irreversible and now the foal has to be born, the long period of waiting is over.

The mare usually lies down at this stage, although very occasionally a mare will deliver her foal standing up. Even now, although as yet there is nothing to be seen, the foal's front legs have been pushed through the mare's pelvis and his head is just entering the pelvic girdle. The uterus is now contracting strongly and regularly, and slowly the foal is being pushed towards the exterior. The next development to be noticed by the observer is the appearance at the mare's vulva of a glistening white membrane, the amnion, covering one front leg which often appears to be upside down. Don't worry, this is quite normal and is possibly connected with some twisting of the shoulders to aid their passage through the pelvis. The other front leg becomes visible in a very few minutes, following slightly behind the first one; again this is quite normal.

Many mares will get to their feet with the foal so far delivered, and it is essential that the mare is not disturbed or distracted in any way: you risk injuring both the mare and her foal if her movements are in any way sudden and result in her becoming off-balance and slipping. She will lie down again very quickly, stretch right out and with many grunts and groans enter the stage where the chest and shoulders of the foal are pushed through her pelvis. The foal is quite a tight fit through the pelvic girdle, and even though the ligaments relax it takes powerful contractions of the uterus to drive this, the biggest part of the foal, through to the exterior. This is obviously the most strenuous part of the birth, marked by vigorous contractions of the uterus, sweating, grunting and groaning.

Now is the moment to slip quietly into the box without disturbing the mare in any way and make yourself insignificant in a corner. By this time two front legs and a nose should be visible. If they are not, just check with your hand that all is indeed well, that two front legs and a nose laid neatly between them can be felt. If they cannot, go and phone your vet and ask him to come very urgently as you think the foal is in the wrong position. A mispresentation is something no amateur can start to cope with and is beyond the capabilities of all but the most experienced stud groom or veterinary surgeon. Such problems are however very rare in the horse; the vast majority of foalings occur quite naturally and the chances are that whilst you squat silent and fascinated in your little corner, your foal will arrive without any help other than the contractions of the uterus.

The nose of the foal usually appears and disappears several times before the neck makes its appearance. Several really powerful contractions mark the expulsion of the bulky parts of the foal and suddenly, with a rush, he slithers into the world. The mare's part in the birth of her foal is over; the hind legs are never expelled by the mother's efforts.

At this stage, the foal will still be enclosed in the amnion and it is a good idea for you to move ever so quietly, so as not to disturb either mare or foal, and just rupture the membrane; then, as the foal takes his first breaths, his head is clear of the birth liquids and there is no danger of him inhaling any of this liquid and damaging his lungs. It is important that this membrane is not ruptured until the foal is born, because if it was done when the chest was still in the mare's pelvis, the foal could start trying to breathe too soon; if he did so whilst his chest was still in the confined space of the pelvic girdle, there would be a risk of crush injuries to his ribs and lungs.

Almost at once the foal will take two or three gasping breaths and then start the rhythm of breathing that will continue twenty-four hours a day until the day he dies. At the same time he will arch his neck, stretch out his front legs, roll over on to his brisket, and rest in this position for some minutes. At this stage the foal's hind legs will still be inside the mare and he will be connected to his dam by the umbilical cord which, you will notice, continues to pulsate for some minutes after the foal is born. I cannot stress too strongly that this cord must *not* be cut or tied in any way. It will eventually be broken by the foal's first struggles to get to his feet.

The reason for not breaking the cord, or tying it off and cutting it, as used to be recommended, is a very important one. At the moment of birth as much as one-third of the foal's total blood volume may be in the placenta, which is still included in his circulation system. This blood will be returned to the foal via the umbilical cord, the ductus arteriosus in the heart closes, and the blood starts to travel round the lungs instead of through the placenta. If the cord is ruptured by interfering hands before this blood has returned to the foal there will be a considerable loss of blood and the foal will start life in a deprived condition.

If the cord is allowed to break naturally it always does so at the same place, about 5cm from the foal's navel. This is the natural break point. The cord very quickly seals itself, preventing any haemorrhage and, very importantly, preventing the entry of any infection. The stump left will be neat and clean with no decomposing length of cord left attached to the foal's navel.

There is no need to dust or spray the navel with any of the recommended medicaments. When the cord breaks naturally, nature's closure will be more effective than any man-made drug in keeping infection at bay. Indeed iodine, so often advised in older books, may be actually harmful and cause irritation and a local cellulitis which may lead to an infected navel and all sorts of problems. If you must use something, use a dusting of a wide-range antibiotic powder obtained from your vet for the purpose.

The mare and foal will rest for some time after the birth, the longer the better, so keep still and do nothing to disturb matters. Do not move about. Do not go out of the box and start chattering. The foal will usually be the first to move, and during his first struggles his hind legs will be withdrawn from the mare, the cord will be broken and you can discover whether your foal is a colt or filly and notice with a thrill that perhaps he has inherited his dam's markings and his sire's beautiful head. The mare will nicker to her foal, and at this time it is a good idea just to pull the foal round to the mare's head so that she can lick him. This licking dries the foal and stimulates his circulation and helps to cement the mother/child bond.

Eventually the mare will get to her feet and the foal's attempts to get to his will become more strenuous. His first efforts to use his legs will result in some spectacular tumbles; he will fall over again and again. Do not assist him to gain his feet: his struggles are essential exercise, which will help him to expand his lungs properly and get his circulation going. Foals who do not get this exercise are more likely to succumb to respiratory disease at an early age. Once the mare is on her feet and the foal is making determined attempts to get on his, it is a good idea to remove the water bucket from the box before the foal falls in it, possibly hurting himself, soaking you and upsetting the mare. It is a good idea to take yourself out of the box at the same moment and continue your observations from over the half-door.

The afterbirth will be hanging from the mare's vulva and is normally expelled about an hour after the foal is born. Should it be so long that the mare is treading on it, it must be carefully tied up into a ball out of the way. It must not be pulled or shortened in any way, as the full weight of the membranes is needed to aid the separation from the uterine wall, and pulling upon the membranes could cause serious injury to the mare. Very occasionally a maiden mare will become upset by the placenta swinging against her hind legs. It does not happen often, but if it does you must get help to hold and calm the mare whilst you tie the offending article up high enough to be clear of her hocks. If left to dangle she could injure the foal when she kicks at the afterbirth and the upset could cause a dangerous rise in her own blood pressure that could lead to an internal haemorrhage. Never use a twitch on a newly foaled mare for the same reason. (To my mind twitches are out of place on any modern establishment. We certainly do not have one on the stud.)

Once you have satisfied yourself that your mare knows her foal is there and is not showing any dislike of him—if there is going to be trouble it usually occurs within about ten minutes of the mare getting to her feet or else not until the foal attempts to feed—you can go indoors and make the mare a warm feed and yourself a cup of tea or coffee. Don't wake the rest of the household; the fewer people about the better at this time. The mare's feed should consist of two scoops of bran, two handfuls of bruised oats damped with enough hot water to make it pleasantly mushy and warm (not sloppy), with a tablespoonful of salt, and two or three tablespoonfuls of black treacle can be added to advantage. Dissolve them in some hot water and add to the mash. You will find that your mare is ravenously hungry and she will probably ignore her foal whilst she eats her feed.

The foal should be on his feet and looking for his dam's udder and his first feed within one hour of birth. He will start searching for his milk-bar in everywhere but the right place, sucking his mother's knees, hocks and stifle until eventually he finds the udder and gradually the teat and has a good suck. We never leave a new-born foal until we have seen him feed and are quite convinced that the mare is accepting her foal happily and completely. One should not assist the foal to get his first feed unless he is being particularly thick about finding the teat— usually big colt foals—or the mare is ticklish and keeps moving out of the foal's way, and threatening to kick, perhaps squealing loudly, every time the foal touches her udder.

Maiden mares foaling for the first time can sometimes prove very shy of allowing the foal to feed. They cannot make out what this strange spidery thing is doing, poking about around their hind legs. In most cases maternal instinct tells them it is

An ideal type of riding horse: Tarnik, a part-bred Arab stallion of the Polish Malapolska breed

A part-bred Arab 3-year-old by an Arab sire out of a cross-bred pony mare. Subsequently a very successful riding pony

An Arabian stallion of the type likely to sire good Anglo-Arabs and part-breds. Big, with excellent limbs and a good front. A proven performance horse and sire of the young stock illustrated in this book

Performance-tested: Tarnik (seen in the first plate), jumping a clear round in an open hunter trial

precious and that they must not kick at it; the most usual reaction is just to keep moving away and eventually, unless someone intervenes and assists by steadying the mare and talking to her and soothing her whilst the foal has his first feed or two, the foal becomes discouraged and gives up. It is not unknown for foals to die of starvation because fond owners imagine that the foal must be feeding. 'I mean they just do, don't they?' It is no use hoping; you must see that your foal is sucking and keep watch that he does so regularly.

If your mare is jumpy and squealing put a lead rein through the headcollar and murmur sweet nothings in her ear whilst your helper guides the foal to the milk-bar and helps him to have his first feed. Do not shout at the mare or scold her unless she seriously offers to kick. She is puzzled and needs reassurance, not threats. Once the foal has suckled a few times, the problem should solve itself anyway. Mind you, some mares will squeal noisily every time the foal feeds for as long as a fortnight after he is born.

Very occasionally a mare will actively dislike her foal and attack him. If by rare mischance this should happen to you, don't try to cope on your own—it can be exceedingly dangerous. Get help urgently. Catch the mare, remove the foal from the box and keep them separated until you have obtained advice from your vet. Also turn to my section on orphan foals in Chapter 6.

It is essential that the foal has his first feed within three hours of birth. The antibodies that protect the foal from infection are contained in the colostrum, or first milk, and these antibodies are absorbed through the gut wall into the foal's bloodstream. The ability of the gut to absorb antibodies diminishes rapidly after birth, from a maximum one hour after birth to about 50 per cent twelve hours after birth, and the gut wall is closed to the antibodies completely twenty-four hours after birth. So if the foal is to receive maximum protection against the challenge of the various bacteria in his environment, he must receive his dam's colostrum within three hours of birth.

Once the foal has had his first feed and the mare has eaten her mash, and both are warm, dry and comfortable, you may leave them and go and have a few hours' sleep. Replace the water bucket, filled with water with the chill taken off, make sure the straw comes right up to the bottom of the loose-box door so that the draughts from a chilly springtime dawn cannot penetrate (remember though not to go shaking fresh straw about—think of the fungus spores and the mare's still-lax vulva), close the top door and leave mother and baby in peace to get acquainted.

You will realise that your part in the whole proceedings has been that of an observer. This is how it should be. Birth is a natural process and unless things go wrong no help is needed. We are only there to recognise the need for assistance and summon it if and when necessary.

The mare should have cleaned (that is, the remains of the placenta and its membranes, the afterbirth, should have come away) by the following morning. Indeed many mares will clean within about one hour of the foal being born, but if this is not the case and the afterbirth is still being retained twelve hours after foaling, your vet must be notified as he may need to come and remove it. The afterbirth should be checked to see that it is complete, and again if it is not, or you suspect that a little piece has been retained within the mare, ask your vet to look in and advise.

I cannot repeat too often that if you think anything is wrong, call your vet at once. He would much rather be called to find a healthy foal already delivered, or a cleaning duly expelled and awaiting inspection, than have to try to remove an already dead wrongly presented foal from a half-dead mare, or to try and treat a grossly toxaemic mare with a stinking retained afterbirth (and that stage takes about twenty-four hours to reach in the horse). The proverb 'he who hesitates is lost' is particularly true when it comes to foaling mares, and if problems do arise the loss is very likely to be both mare and foal if professional help is not sought very quickly.

One last thing on the subject of the afterbirth: do make sure you dispose of it so that it does not make an unwelcome reappearance aided by your dog or cat—both species seem to have a fascination for the things! So dig a deep hole in the muckheap, and spread lots of fresh manure on top.

Keep an eye on the foal to make sure that he is dunging normally and has passed the meconium or foal dung. Should there be any sign of discomfort or straining, once again your vet should be called to give the foal an enema and a dose of liquid paraffin. Watch both mare and foal for any signs of discomfort. Make sure the dam has plenty to drink, keep all her feeds non-heating and damp, give her as much good hay as she wants (don't feed it in a net though; the foal might just get hung up in it), and as long as all is well keep away, leaving mother and child to recover from the stress of birth and get to know one another.

Resist the temptation to spend all day gawping, and even more important keep everyone else from doing so. Peace and quiet are essential for the wellbeing of both mare and foal for the first few days.

There is just one final word of warning. Some mares become very foal-proud, occasionally even to the point of being positively dangerous to their attendants. Our senior matron is like this. About half an hour after she has foaled you are given very clear indications that it is time to leave, and for the next two or three days all you do is poke food and water round the door and retreat quickly. Mares like this make first-class mums but they can be holy terrors to cope with for the first few days after foaling. Leave a headcollar on so that you have something to grab if really necessary, but preferably leave them alone and keep out of the way.

5
The First Two Weeks

Once your long-awaited foal has made his or her safe entry into the world, you face the period when risks to both mare and foal are greater than at any other time, except possibly at the moment of birth. However a very large percentage of the problems that occur are entirely avoidable, most upsets during this time being caused by lack of cleanliness, thoughtlessness and too much disturbance.

To take the last point first, a constant stream of visitors, some of whom may not be very knowledgeable and therefore rather noisy, will worry the mare and tend to upset her milk supply. Too much disturbance could even cause her to reject her foal. A newly foaled mare is very protective towards her off-spring and no matter how easygoing she may appear to be, a non-stop procession of 'gawpers' will put her on her guard. They will make her irritable and tense, just when she needs to relax and recover from foaling and adjust to her new duties as a brood mare. Likewise the foal at this stage needs all the rest he can get; jumping about and showing off can come later.

So, for the first three days no visitors—even the family are best advised to keep away except for essential visits to feed, water, hay and observe mare and foal to check that everything is well. You will notice that I do not mention mucking out. Just spread a little clean straw over any wet patches after removing the worst of the droppings into a skip. This clean straw should be shaken up outside the box and taken inside in a carrying sack. Shaking up straw in the box is to be avoided for the first few days as fungus spores are released in quantity from any straw, no matter how clean, and these spores can find their way into the mare's still-lax uterus, possibly setting up an infection and causing subsequent infertility.

The introduction of stable tools into a confined space with a wobbly foal and possessive mum is courting disaster; anyway, the deeper the bed the better for very young foals, as it provides insulation from damp chilly floors and draughts. By a deep bed I do not however mean a mountain of loose straw to get tangled round the legs of the foal.

Going back to the dangers of stable tools in boxes, I well remember visiting an apparently well-run stud, early one rather cold spring and watching with horror as the owner proceeded to enter a very small box (10 x 8ft) armed with muck fork, shovel and broom. The box housed a valuable show-pony brood mare and her week-old foal; they had been unable to go out due to bad weather. The foal was a very lively colt who, stirred up by the activity, started to buck and cavort around the box, his dancing legs just missing the razor-sharp prongs of the muck fork. The mare, meanwhile, had become very upset and started rushing round and round the box after the foal, slipping on the floor with every stride, eventually knocking the broom and shovel to the floor, from where they had been propped against the wall.

Unable to bear it any longer, I went into the box, caught the foal and held him in a corner, whereupon the mare stopped rushing about, and peace and comparative safety returned. In fact, just a little forethought would have prevented the situation occurring in the first place. Less than 25 yards away was a large, empty, covered school—how much better to have turned mare and offspring into the school for an hour's exercise while the box was cleaned! The foal could have stretched his legs, the mare had a roll and the mucking out been completed without the dangerous incident I witnessed ever happening. As I have said before, 95 per cent of all accidents are caused by thoughtlessness and are avoidable.

Watch Mare and Foal Carefully

You mare will almost certainly develop an enormous thirst after foaling as she comes into full milk. It is *essential* that an adequate supply of clean fresh water is kept with her at all times. Two big buckets must be kept filled and with her twenty-four hours a day. A big mare in full milk can drink upwards of twelve gallons of water a day, and if this is not available it is bad for both the mare and her milk supply.

For the first two or three days after foaling both the mare and foal should be closely observed for any abnormal symptoms. Should the mare show any signs of unease, sweating, loss of appetite, a dirty smelly discharge from the vaginal opening, or uninterest in the foal (do remember that non-accepting mares *don't like* their foals) your vet must be told instantly, with an urgent request to attend. Such alarms are fairly unusual, but it is well to be on your guard and be much more observant than usual. Noticing any symptom in the early stages and treating it promptly means the chance of recovery is very much greater.

The foal too must be closely observed at regular intervals to check that he is feeding regularly and dunging normally. Many first-time breeders worry unnecessarily because the mare does not seem to have much milk in her udder; how can she, if the foal has drunk it? The best guide to whether a foal is getting sufficient milk is his general condition, the length and colour of his urine streams and the length of the intervals at which he feeds.

A satisfied foal is rounded and shiny and has good feeds at regular intervals, with periods of sleep or play in between, whereas a hungry one looks shrunken, his coat stares, and he will be forever poking around his mother's udder, sucking so hard that the mare may well become sore and peevish. Nor will he rest as he should.

The best guide of all is to observe how the foal urinates. There should be a good stream of virtually colourless liquid that lasts for a significant length of time. Little dribbles of concentrated urine almost certainly mean that the foal's liquid intake is insufficient. This happening is fortunately fairly rare, but if it does occur as much as possible must be done to encourage the mare's milk supply. She must be fed damp feeds, with sliced apples and carrots, succulent grass must be cut for her, anything in fact that will increase her liquid intake and tend to increase the flow from the milk-bar.

The foal, too, can be persuaded to nibble a little of mother's feed particularly if her rations are split into two and fed from two bowls on the floor. Do not worry about her not getting the full amount. She will eat both lots anyway, but it is surprising how a very young foal will soon learn to eat solid food, particularly if he is hungry; if the mare's milk supply is really inadequate it can be supplemented by giving the foal little feeds with milk powder added. In this case it is a good idea to put the foal's feed in a bowl and hold it for him.

It is not a good idea to try to supplement the mare's milk supply by bottle feeding, unless it is absolutely essential for the foal's survival. I say this because it is all too easy to give the foal a bottle, and he then finds it easier to get his feeds that way and does not bother as much with mother's udder; the lack of stimulation tends to dry her off, thus making the situation far worse than it was originally.

A very full udder should also be regarded as a danger sign; as already said, healthy foals do not leave milk to run to waste. If the bag is overfull with milk dripping away, it means that the foal is not feeding, and a young animal that is not feeding is a sick animal; loss of appetite is the first symptom to show in most cases of scour or obstruction, so call your vet.

The foal's dunging must also be observed. The dung should be soft but formed and passed regularly without any sign of looseness, straining, abdominal pain or discomfort of any kind. Sometimes a small portion of meconium or foal dung is retained and caused a partial obstruction. However once again, should there be any sign of either scouring or obstruction, call your vet urgently. A young foal can dehydrate beyond the point of no return in a matter of hours if abdominal upsets are not treated promptly.

This is perhaps a good place to mention that about 7 to 10 days after foaling, the mare will have her foal heat. This heat should be ignored for breeding purposes and the mare should not be covered then, for reasons that I will explain elsewhere. However certain substances are secreted in the mare's milk at this time which may cause a foal to have loose motions. This scour is not infectious and will clear up when the mare goes out of season, but it must be watched very carefully and any sign that the foal is being weakened must be reported again to your vet, who will probably prescribe a simple kaolin mixture. The foal's quarters must be kept clean too. (See Chapter 6 on the care of a sick foal.)

The mare's feeding for the first week or so should be directed towards keeping the milk-bar in full production and being easily digested. Three or four small feeds a day are better than two big ones; bran, oats, etc, should be fed damp, with the addition of sliced apples, carrots, etc—in other words, traditional feeding. I do not recommend proprietary nuts of any kind at this time; they are thirst-making and some may contain additives to help condition and bloom. I am not convinced that these additives are always harmless, so I prefer to avoid made-up feeds which may contain them.

However, having given all these warnings, the chances are that your foal will turn out to be a normal young horse, who drinks, sleeps, gets up, has a stretch and yawn, drinks some more and goes back to sleep again, lying flat in the straw, eyes tight shut and sometimes even snoring.

Your foal should be ready for his introduction to the big, wide world by the time he is two or three days old. It is not a good thing to leave mother and baby in the stable for too many days after foaling; but before they do go out for the first time, both must be physically well, the mare must have accepted her foal completely, the foal must be feeding regularly without help, and most importantly, the day should be dry and warm. Also you must have help, at least one other person, preferably two, bearing in mind that one quiet, patient, competent assistant is worth twenty rush-and-tear merchants!

The mare should be fed as usual in the morning but make her feed somewhat smaller than normal so that she will go out feeling slightly hungry and settle quickly to grazing. While she eats her breakfast, go out into the field they are to use and check thoroughly for anything that could harm a foal enjoying its first taste of freedom. Remember the mare knows the pole by the gate sticks out a bit and that there is an oil drum lying in the grass where the children left it after their last jumping session; the foal does not, and may well injure itself in its first exuberant dashing about if such hazards are not removed.

Ideally the field used should be small—about 1 acre is ideal—and have been free of stock for some six to eight weeks to give fresh, clean grass time to grow. It is also easier if the field is not very far from the stables, as you will find there is nothing so strong, awkward and generally difficult as a three- or four-day-old foal who doesn't know what is going on anyway, and the less distance you have to push-carry-stagger with him the better.

It of course goes without saying that the chosen field must be free of barbed wire, unfenced ditches, ponds, boggy areas and similar hazards. The mare and foal must have this field to themselves at least for the first week. Other stock, particularly horses and cattle, can chase and harm a new foal. I know of at least two cases where bullocks have trampled a young foal to death. Any horses in neighbouring fields, with intercommunicating gates, are better brought into their boxes. One can well manage without a battle between a possessive mum and an admiring auntie over a metal gate, with foal possibly caught in the cross-fire.

One last thing before fetching mare and foal: if you want to take some photos of the foal's antics, fetch your camera (checking to see that the film has been put in!) and hang it on the fence or put it safely in the hedge, where it is ready for action the moment the foal is released. It is frustrating in the extreme to dash indoors for the camera, leaving a cavorting foal who was providing many enchanting picture opportunities, only to return after finding the film and loading it to see the mare quietly grazing and the foal stretched out asleep in the sun!

Now let us consider the more serious business of getting mother and child from stable to paddock. Naturally the mare will still be wearing a headcollar after foaling, so the mare's usual handler goes quietly into the box and slips a lead rein through the headcollar. Do not buckle it on but hold both ends firmly so that if the foal gets its head hung up in the lead rein, it will fall free if one end is released. Actually a ten-foot length of soft rope with no fastening is best of all, but this may not be available and a lead rein works well. A halter rope is not long enough and the clip on the end can cause injuries.

Once the mare is secured, one of the two helpers goes into the box and catches the foal by placing one arm round his chest and another round his quarters and then looping a length of soft rope or stable rubber round his neck. Do not make a grab for the foal; work up to him gently. Obviously the rope must not be tied or looped in any way that could pull tight. All ropes, lead reins, etc, must fall free should either mare or foal get loose; some quite horrific accidents have occurred through trailing ropes. It may be easier to catch the foal if he is restrained in a corner, by the mare being placed across the corner with foal and you behind her; mother's presence is comforting and her bulk makes a satisfactory soft padded barrier. Having caught the foal, the other foal handler comes in and links hands fore and aft with the first, thus making the foal a mobile human baby-walker. The mare is then led quietly out and turned to face the box door so that she can see that her baby is following.

It is essential for safety reasons to take the mare out of the box first. For one thing she has been cooped up for three or four days and is usually keen to get out. Her going first prevents her becoming suspicious and irritated and rushing out in hot pursuit of her baby whom, she is convinced, no matter how well she knows you, you are trying to steal. It is very easy for the mare to injure her foal, her handlers and herself by charging through the door, so take her out first. After all, in nature the young follow their dams closely for some days after birth. They do not lead the way.

The foal is then gently persuaded to follow his dam. Remember everything outside the box will be strange and possibly frightening, and he may not be very eager to leave the security of the stable. You may well end up more or less carrying him to the field, but as long as arms remain linked around him you have the situation under control. We do not, by the way, fit a foal-slip at this stage. It is quite unnecessary and can prove dangerous. Once everyone is outside, the mare is led quietly and slowly in front of the foal to the field. The mare leader should remember that the foal party may not be making very rapid progress and must not go rushing off.

This routine is a safe and trouble-free one in most cases. But if the mare is very foal-proud, it is safer for the mare leader to take mother outside the box, and for both foal handlers to go *instantly* in and shut the door behind them and catch the foal. The mare is allowed to watch proceedings over the door to see her foal is still there. In this case the foal handlers must be as quick as possible. Once the foal is secured, the mare handler opens the door for the others and leads the mare firmly away.

Once through the field gate, the foal is taken well out into the middle of the field and when the mare handler has shut the gate *and said so*, the foal can be released. The mare should have been turned and released facing the foal, so that she can see where he is and does not knock him over in her anxiety to find him. Leave the headcollar on the mare.

The foal will take a moment or two to find that his legs are for cantering and bucking on, but then both will probably go for a short spin round the field, the mare usually going down for a good roll afterwards. Both things are excellent, the exercise helping to dislodge any debris remaining in the uterus after foaling and helping it to contract. Keep half an eye open to make sure the mare does not roll on the foal or kick it when rolling; most mares are very careful but there are exceptions. Keep watching—as if I needed to tell you to do so—until the mare has settled to graze.

The next job is to give the mare's box its post-foaling clean out. Take all bedding out of the box, scrub the floor with cold water and the yard broom, adding a weak disinfectant to the final rinse. Leave the doors open and the floor drying while you have that cup of coffee you have been dying for. The box should be rebedded with a clean deep bed of fresh straw well into the corners and up to the door as before.

Do not leave mare and foal out in the field for very long this first time. Two or three hours is more than enough. When you go to catch mare and foal, go with two people to help as before, armed with some feed. It is the one time we find we may need it, as the mare will probably tend to keep foal and herself out of your way, particularly if she is foal-proud; but most mares will usually eventually come for some food. It is a good idea if the foal-catchers wait out of sight till the mare is caught; the mare may well feel threatened by three humans advancing on her. If very, very foal-proud there is the odd mare who will come for you rather than be caught. If you are faced with this situation, rather than have a dangerous upset leave her until she feels differently. One particularly foal-proud mare of ours went out for a few hours with a new foal and it was a week before we could get near her. She was fed in the field of course, and the foal was a big, strong filly who could well cope with the situation. This, however, is fairly rare.

One foal handler catches the foal as in the box by gently driving the foal (no rushing about, no waving arms, no shouting; walking and soft noises only) into a pocket made by holding the mare at an angle to the hedge or gate, head to fence, quarters at an angle of 45 degrees to the gate, with the mare handler being most particular that the foal does not get caught in the gate. The foal handler then quietly catches the foal, the second handler forming the walking cradle as before, and when everything is under control, in you go. (A gate that opens outwards is a great help here but that is not essential.)

The best way of catching a foal is not to make a grab for its head end—that is laying the foundations for head shyness and difficult catching later. Start by scratching him on the top of his rump, talking sweet nonsense — most foals love this — and gradually work forward until you can gently slip the rope round his neck and your arm round his chest. Remember your foal has no reason to be scared of you; all young things like to be firmly, gently and confidently handled — it makes them feel secure.

We never let our foals run behind the mare. It is bad discipline; the foals do not get handled or learn to lead, and as they get stronger and more independent they have a habit of going off exploring, which can result in all sorts of crises. Two handlers for the foal will only be needed for a day or two. The foal will soon learn to lead from the soft rope looped around his neck and a hand on his quarters just to help control him.

Once mare and foal are back in their box, give mum a little corn feed, then go away and leave them in peace. The foal will very shortly be sound asleep on his lovely deep bed recovering from the exertion and excitement of his first day out.

The daily outings can gradually be lengthened until by the end of the first fortnight, mare and foal can, if the weather is suitable, be left out all the time, being fed in the field if grass is not in plentiful supply or of good quality. One last thing: for the first fortnight we do not like to let our foals get wet. A young foal getting really cold, wet and tired, (they will not lie down on wet ground) is asking for trouble, in the shape of chills, scouring and pneumonia. So for the first fortnight, grab your mare and foal before the washing!

One other word of advice for family breeders. Do emphasize to all the children that foals are not playthings. They must have peace and quiet and be treated with the same sensible respect as one gives to any horse.

6
The Care of a Sick Foal

No matter how careful you are there is always the chance that your foal may become ill and need nursing. A cold, the snuffles, an upset tum when mum comes in season, all these things can cause problems to a very young foal.

The purpose of this chapter is not to discuss the treatment of disease; that is the vet's job and I cannot stress too strongly that anyone, no matter how experienced, who has a sick animal should call in professional assistance as soon as trouble is detected. It is far easier for the vet to treat the symptoms he sees promptly and correctly, with a good chance of success, if the owner has not already treated the animal with Grannie's Colic Drench and the baby's gripe-water mixed, thus possibly allaying the scouring temporarily, but leaving the infection causing the trouble to continue unchecked.

I use scouring as an example but the principle is the same with any infection. Even more dangerous is the temptation to a farming family who just happen to have a dose of penicillin handy. They may inject this regardless of the symptoms, probably in insufficient quantity, with no knowledge as to whether it will deal with the problem. There is a serious risk that such an action will set up an antibiotic resistance, making the infection incurable. So PLEASE call your vet and let him treat the animal properly. Your job starts when, having looked at the foal, the vet smiles sweetly at you, hands you a large bottle and says 'Just give him 10ml of that 4 times a day' and departs in a swirl of dust.

Actually your job does not start then; it starts as soon as you notice the animal is unwell. Good nursing care, combined with prompt appropriate treatment, has cured many apparently hopeless cases.

Keep the Foal Warm

The first thing to remember is that a sick animal feels the cold and needs extra warmth: a good deep bed right up to the door of the box, the top doors closed if it is at all chilly (and most spring days are), and the foal itself kept warm by use of a rug. There are expensive foal rugs on the market which are fine for a healthy foal travelling, but these are useless for a sick animal; they may become soiled and unhygienic. Much better is a disposable rug which can be quickly and easily made, and burnt when it needs replacing.

The method is as follows. Obtain a medium-sized thin hessian sack and cut a sheet of kitchen foil to fit inside the bottom half of the bag; place this in position and on top of that put a *thin* layer of hay, then fold the top half of the bag over so that there are three layers of bag under the foil. You need an assistant at this stage to hold and support the foal. It is sensible if the mare is either held or tied up at this point too, as she may not approve of a hay-lined baby! The rug is then placed in position over the foal and held by two strips of old bandage, sheet or anything soft. Hay cord is not satisfactory; it rubs and does not give. The

The emergency rug on the foal, secured by strips of old bandage tied through the sack

fastenings are adjusted to the size of the foal by making holes in the hessian bag (careful here) and they are then tied through. The rug is secured round the foal's middle by another piece of bandage. See the diagram.

This rug, when finished with or dirty, can be burnt and a new one made if needed. Obviously you need to watch it to make sure it stays in position, but you will be attending at frequent intervals anyway so that is no problem. This rug retains all possible bodily heat and is always readily available to hand which is more than can be said of proprietary foal rugs. I do not like infra-red lamps as a source of heat because of the fire hazard. The rug is quick and easy to make and can be on the foal in a matter of minutes; any time saved is points won in the battle for the foal's survival.

The Foal's Comfort and Cleanliness

The other thing is the bodily comfort of the animal. If it is scouring, its quarters and hind legs are going to become very dirty and caked with excreta. If left, this will form an impervious mat with the hair; air will be excluded and the skin will slough off, leaving some areas that may take months to heal and may even scar. So at regular intervals take a bowl of warm water with some mild disinfectant (Savlon is ideal) and lots of old sheet or kitchen roll, wash, dry and leave clean all those parts affected. A dusting of baby powder to dry off the skin does no harm either.

Should the foal have a dirty nose or discharging eye the same procedure should be followed, of course dispensing with the baby powder in these cases.

Do remember though that cleanliness is of the utmost importance. Hands should be thoroughly washed before and after each washing session, and the same bowl or bucket kept separate from the rest for the washing water. Dirty water should be tipped straight down the drain and all soiled cloths or dressings burnt. One cannot be too careful, particularly if there are young children in the household or other foals around.

(*above*) A mobile baby-walker for a young foal; (*left*) leading a young foal at a show. Note the lead rein looped around the foal's neck, giving close control

Ready to wean: a normal healthy young horse. A 6-month-old foal at a show with her dam. She is now a champion endurance horse

Growing up. The foal seen above now grown into a nice yearling, smartly turned out to compete at a major show

One other thing. Should your foal go down and not be able to stand, do make sure that there is a soft clean cloth under its head to prevent the prickly ends of straw injuring its eye.

Make sure it is turned from side to side several times a day. Pneumonia caused by compression of the lungs can easily occur if the foal lies on the same side without moving for many hours. Pressure sores may also be caused. Again it is easy enough to turn a sick foal once you know how. Make sure that his front legs are stretched out in front of him. Tuck up his hind legs close to his body, roll him over on to his brisket and so over on to the other side. If he is a big foal, you will probably need help to do this. Once he is resting on his other side, gently massage, with a slow circular movement of your hand, all those parts in which the blood circulation is likely to have been impaired by pressure. Hips, rib cage and shoulders are the main areas that need constant attention.

Dosing the Foal

There remains one major problem. The vet has seen the foal, reassured you that the medicine he has prescribed will soon have the foal's scour or constipation cured, left you with a large bottle and rushed off—always to a calving or milk fever, so that there is no possible chance of his helping you to transfer 10 ml of that revolting-looking mixture from the bottle to the inside of the foal. I will guarantee that unless someone tells you the easy way to give a foal medicine, thirty minutes later most of the medicine will be over you, the foal, the mare, the stable and the bedding, but certainly not inside the foal.

There is an easy and simple way. Ask your vet before he escapes for an old 10 or 20cc unbreakable plastic hypodermic syringe (no needle on it of course). Fill the syringe with the required dose. Back the foal up into the corner of the box, leaving the full syringe where you can reach it. Hold the foal in the corner with your body, and with one hand gently lift the foal's head to the horizontal while the other hand picks up the syringe and slides it into the corner of the foal's mouth. Still keeping the head horizontal, slowly depress the plunger of the syringe so that the medicine flows in a slow trickle to the back of the foal's tongue where the normal swallowing reflex takes over and the medicine slides down without fuss.

Be gentle and slow; there is no need to ram the syringe into the foal's mouth and eject the medicine in a forceful stream. This could be dangerous and lead to choking. The syringes supplied with worming pastes will suffice for this job, but are not ideal as they often leak up the plunger and the net result is that the medicine ends up in your eye instead of down the foal's throat.

Cuts and Scratches

The other things you may have to deal with are minor cuts and scratches. Obviously if any injury is serious, ie deep or extensive, the vet must be summoned promptly to stitch and clean if necessary; punctured wounds should be treated with particular suspicion. Much more common is the odd cut sustained who knows how, not serious enough for professional attention but sore, messy, and in need of cleaning up. Ah yes, we just bathe it off with a little warm water and disinfectant, don't we? Have you ever tried this seemingly simple exercise? The foal is frightened, the cut hurts and there are you attacking him with that foul-smelling liquid! Again chaos reigns.

There is an easy way. It is helpful but not essential if an assistant can steady the foal, arms round quarters and forehand, as the foal gains confidence from being held close. Once more the old hypodermic makes the task relatively trouble-free. Fill the syringe with a solution of dilute hydrogen peroxide. (Dilute 1 part of hydrogen peroxide with 3 parts previously boiled and cooled water.) Trickle this solution out of the syringe so that it runs down over and into the injury. Do not try and poke the syringe into the cut; the bubbles that are released will gently clean the injured tissues, and they will lift away all the debris and dirt and remove any messy discharge or dried blood. Keep gently flushing until the wound appears clean. This treatment should be repeated until the injury is obviously healing satisfactorily.

Hydrogen peroxide is ideal for cleaning and disinfecting minor wounds, as even though it generates a fearsome amount of froth its use is both painless and odourless. It contains nothing more than water and oxygen; it is a naturally unstable compound which breaks down into its component parts as soon as it comes into contact with any organic material, the spare oxygen escaping as bubbles, which carry away all extraneous matter with them, and of course the free oxygen acts to give the injured cells a boost and help the healing process. Hydrogen peroxide is particularly desirable for cleaning ragged, dirty or punctured wounds. Clostridial bacteria, and they are the ones that cause tetanus, gas-gangrene and other similar horrors, cannot survive in the presence of oxygen and so hydrogen peroxide acts as a double insurance against infections of that sort. Of course any injury should always be regarded as a possible cause of tetanus in the horse and appropriate preventative measures taken in the shape of vaccination.

Attention—and the Vet

Long complicated instructions and horrifying descriptions of various symptoms and their treatment are out of place here. I cannot stress too much that the restoring to health of any sick creature depends firstly on correct early veterinary treatment and simple caring attention. What I mean is summed up by this little story: a famous physician was touring a hospital ward with his retinue of students. He stopped at the bed of a tired, ill-looking woman, picked up her notes and read 'Admitted last evening, nervous exhaustion and stress.' 'RWN and TLC indicated here, I think,' he wrote, and passed on to the next case. The students were too much in awe of him to ask what the letters meant. A fortnight later the same physician and the same students passed on their rounds again. The same lady was now sitting up in bed looking rosy-cheeked and happy. 'Ah, Mrs X, I think you can go home now,' he announced, and turning to the students he said 'That treatment is always successful.' One student ventured to enquire, 'Sir, what do those letters mean? What was the treatment you prescribed?' The great man smiled and said, 'Ah, yes—well, RWN—rest, warmth and nourishment, and TLC—tender loving care. Together they effect more cures than all man's medicine.'

If Foal or Mare Should Die

It would be nice if the end of this chapter need not be written, but there are two disaster situations that should be mentioned. Should you by great misfortune lose your foal from any cause, do not rush to remove the body. Leave it with the mare until she no longer frets over it, usually about 2-3 days. Then take the mare out of the box and turn her out with company. Then, and only then, you can ask kennels to come for the body.

The other great misfortune that does just occasionally happen is that you lose your mare. Should this happen you do have problems, but do not panic. Do not be tempted to destroy the foal, as I have known vets advise, rather than help cope with the problem. This is not the place to discuss the rearing of an orphan foal in detail but should you ever be faced with this disaster, the following points may well tide you over until a satisfactory solution to your problem can be found.

Should the foal not have had its colostrum, eg if the mare dies foaling, it *must* be put on a broad-spectrum long-acting antibiotic cover, given intramuscularly, at once. This cover must be maintained for at least a fortnight until the foal has had a chance to make its own antibodies against disease.

Do realise the foal's life is in danger only from lack of food. Whatever it was killed the mare—haemorrhage, birth injuries, colic, etc—cannot affect the foal. He was a separate living entity the moment the cord severed, only dependent on his mother for food, care and warmth.

All young things will take food from a bottle. A baby's bottle with a human teat is the best for a foal, making sure the orifice is enlarged and the teat securely fixed on the bottle. Should you have difficulty getting the foal to suck or not have a bottle, start the process going with our faithful friend, the old hypodermic syringe with no needle. Fill it with milk, and dribble it into the foal's mouth. A hungry foal will soon come back for more when warm milk makes its way to his tum. Obviously all feeds must be given at blood heat, slightly cooler being better than too hot.

Little and often is the secret of success. A half-pint feed every two hours will cope for starters. Remember, though, that absolute cleanliness must be observed. All utensils, feeding bottles, etc, must be *boiled* between feeds. And 'every two hours' does mean night and day.

The foal must be kept warm. My foal rug will help here.

There are mare's-milk replacers on the market, but the supplier is nearly always on the other side of the county and appears to be ex-directory. In truth the solution is closer to hand than that. Goat's milk is excellent for rearing foals and can be fed undiluted and without additions. It can be easily obtained from most health-food shops. But remember that in an emergency cow's milk diluted with one-third boiled water, and a tablespoonful of glucose added to every feed, will keep a foal alive until the shops open in the morning. It is no good being so particular over what you give the foal that he is dead from hunger by the time the special milk arrives.

As soon as the hour is reasonable, and do remember no one is likely to be at their most helpful if awakened from their bed, seek help urgently from the National Foaling Bank, a forward-thinking local breeder or anyone you know who has successfully reared an orphan foal. What you need is a foster-mum and the best foster-mum I know is a large milky goat. The foal will soon learn to suckle straight from the goat if she is stood on an old table or pile of bales.

The goat provides company too, and this is useful as foals brought up by humans only tend to regard themselves as human, and life is not made easy when a large four-year-old makes friends with someone he considers as one of his own kind by putting an arm around his or her shoulders! Your foal must learn that he is a horse and subject to horsy discipline, and this is much easier if a foster-mum is provided. We all know how much of a nuisance tame lambs can be.

If all fails and you can't find anyone to help you with the details or find a foster-mum, put an appeal out, over either the local radio or local TV. They are very good and will always help if they can.

Do not be tempted to try and rear a foal on the bottle. For one thing you will always have this tame-lamb problem, and secondly the amount of work is killing, unless there are five or six of you to take turns with night feeds—and the 'every two hours' bit has to go on for at least 8 weeks. No one in the position of the average amateur breeder can cope without having proper rest for that length of time; it is not fair to oneself or one's family.

7
Babyhood

Once your foal has passed the tiny infant stage, life becomes much simpler. A routine as to turning out, feeding, etc, will quickly establish itself and eventually, as the days grow longer and warmer, your mare and foal can be left out all the time.

Remember that they must be seen at least three times a day—early morning, midday and evening—then if there is any slight injury, or the first warning symptoms of the foal being off-colour, it can be attended to before anything more serious develops.

If the weather happens to turn cold or, worse, very wet, your mare and foal are better brought into their box, and in the case of very heavy rain left there until it stops, fed and watered of course. Foals will not lie down in the rain or on waterlogged ground and they very soon become tired; a tired foal chills easily, and there you have the first steps to pneumonia. It is also as well to remember that a foal's coat is much more woolly and less waterproof than that of an adult horse and soon becomes soaked through. So, again, if the rain is heavy and seems likely to last more than half an hour or so, bring the pair of them into a well-bedded box.

One other point whilst on the subject of weather. If it looks like thundering, it is always worth bringing in your mare and foal. I do not like horses out in a thunderstorm, because apart from the very slight risk of them being struck by lightning, there is the much greater risk that they may become panicky and rush into things, particularly if the storm is during the hours of darkness. So if it looks and feels thundery bring them in to their box and leave them there. It does not matter if you have not got any hay, give them a feed and leave them. They will not starve to death in six or eight hours.

One word of caution here; if ever for any reason you have to go about your business round horses with a torch, do not flash it about looking for them or use it in such a way as to make alarming shapes on stable walls, etc. Horses in general are terrified of flashing lights and grotesque shadows, so if at any time it is essential to use a torch do so very carefully, talking all the time so that your animals know you are there. You may think I am being ultra-careful but I personally know of three instances where horses have been killed after being panicked by thoughtless use of a hand lantern.

Will you Breed from your Mare again?

The next major decision is whether you wish to cover your mare again. If you want to breed a foal every year, she must be returned to the stud of your choice some three weeks after foaling, so that she may be covered on her first proper season after foaling which is usually between 28 and 32 days after the foal is born.

Do not be tempted to send the mare back to stud for the so-called foal-heat. There are many reasons against this. Firstly, if there has been any internal bruising of the mare or any very slight infection, service at seven or nine days after foaling is going to aggravate the matter, whereas in a month nature will have healed the bruising and may well have cleared any slight infection. It is a fact that only 20 per cent of mares covered at the foal-heat hold to service. If you are in any way doubtful about your mare having any infection, it is always worth getting your vet to take a swab before she goes to stud and get the trouble cleared so that she returns to stud clean and ready for service.

Secondly, a mare with a five- or six-day-old foal is still at the possessive foal-proud stage and may well be aggressive towards other mares, so when she goes to stud and gets mixed with other mares and foals, kicks and all sorts of injuries can easily occur. Our mares and foals always spend their first week to ten days on their own anyway.

There is also the stress of travelling a very young foal, and the risk of a foal picking up infections from other horses before he has had a chance to build up any resistance.

One final point is that on a busy stud, a foal-proud mare who may not wish to be caught by strangers is a nuisance and is unlikely to get brought in out of the wet, etc, even if the stud does have time or the room to bring everything in. A month-old foal will be better equipped to fend for itself.

Foal-proud mares are not easy to cover, and if your mare is returning for service, 24 days after foaling is about right.

The only snag about sending your mare back to stud is that your foal goes away at about three weeks old and may remain away for a month or more, during which time he will grow and develop to such an extent that he will be barely recognisable on his return home. It seems to me that if you want to enjoy your foal growing up and to make sure his handling and education progress as you personally would wish, it is not a bad idea to cover the mare only every other year, so that she goes to stud without a foal at foot. No stud, no matter how efficient, can possibly spare the time to educate and handle everyone's foals; anyway why should they? The stud fee does not include halter-breaking and leading foals, and on a busy stud at the busiest season of the year there simply is not time, even if there were the inclination. It is surprising how wild a foal can become if not handled for a month or two, even if it has been led during its first days.

Leading the Foal

At some stage your foal must learn to lead from a foal slip. We do not lead foals under a month old from a slip, but always from a soft rope looped around the neck, never tied but always held in such a manner that should the foal get loose the rope falls away with nothing to frighten or tangle.

When it comes to haltering, we use the following technique. Your foal is obviously used to being handled and led, having been led to and from its stable, so it is no problem to catch it in the box. On the appointed day, arrange some quiet assistance. The helper puts a headcollar on the mare and then when the foal handler has caught the foal, again with the soft rope looped around its neck, the mare handler pens both the foal and handler in the corner of the loose box. The foal slip can then be gently introduced. Carefully slip the noseband on and off until the foal no longer objects. This may take two or three days of quiet patience, say 5 or 10 minutes daily. When the foal accepts without fuss the fact of something sliding over its nose, the headpiece may then be fastened. Be careful not to do what so many people do without thought—flip the strap up from the off side and catch the unfortunate animal a smart and, I am sure, painful flick in the eye! Always reach for the headpiece and lift it gently over the foal's head and fasten it quietly and without jerking. This should be repeated until the operation can be carried out without resistance, being careful always that all movements of yours are smooth and jerk-free, and above all confident. Nothing upsets a young animal more than fumbling, shaking, uncertain fingers.

Leading a foal. Notice the equal use of a rope looped round the neck and a foal slip

Leading from the headcollar is taught in a progressive fashion. Once the foal is used to having its slip on when you lead it to the field behind mum, still from the soft rope around the neck, use a rope long enough for the spare end to go through the ring on the headcollar, (see the drawing) and to be held by your spare hand. This means that pressure on the foal slip is introduced gradually. Eventually the loop round the neck can be dispensed with and your foal will lead from the slip without fuss.

Three important warnings though: always use a soft cotton foal slip with an adjustable nose as well as headpiece; never use one made of nylon which does not give and can rub most horribly. We once had a foal sent to stud whose owner had put a new leather slip on it at birth and left it without alteration until the unfortunate infant came to us at three weeks old. The foal had grown, the slip had not, and the resultant ulceration was horrific; it took us two days to cut the foal slip from the foal's head.

So use a soft slip, which must never be left on, either in the box or out in the field. Foals poke about in the most peculiar places and can get hung-up no matter how careful you are. If my leading routine has been followed there will be no trouble putting the slip on the foal and taking it off anyway, and to put the slip on in the field gets the foal into adult habits of being caught without fuss.

The other thing to remember is, never, never to fasten the lead rein through the ring on the headcollar; always use the leadrope double, with the leader holding both ends. Then if your foal decides to play up and gets loose, and *they all do* at some stage, the rope will fall away and leave nothing dangling to frighten and get caught up. Again many horrible and avoidable accidents have occurred through this simple piece of common-sense not being followed.

Perhaps this is a good place to stress that headcollars should never be left on when they are not actually being used to lead the animal, be it mare or foal. This applies as much in the stable as in the field (except for the foaling mare, as described earlier), and is particularly important now that nylon collars are in common use. Nylon does not break, having a breaking strain considerably in excess of any pull that can be exerted by a horse. This means that should your mare or foal be unlucky enough to get caught up by the halter, serious injury or strangulation is a distinct probability. In 90 per cent of cases where animals have become entangled by the headcollar they are dead when found, so make it an absolute rule to remove your mare's headcollar and the foal's slip every time you turn them out or put them in their stable.

I have already made the exception that a soft leather headcollar must be left on a foaling mare. Here the need for grabability outweighs the likelihood of the mare getting caught up, and anyway the mare will be under constant supervision at this time. The same applies when she is first turned out with her new foal.

Mares that are difficult to catch can be a problem and sometimes it is necessary to leave a headcollar on such animals, particularly when they go away to stud and have to be caught by strangers. In this case let the headcollar be an old, soft, leather one which will break rather than strangle.

Avoid Titbits and too much Fussing

Two further points. Your foal should be taught to lead striding out freely beside you, not rushing off in front and not trailing along behind. He should, by the end of the summer, trot on when asked, wait while gates are opened and shut, and generally behave in a civilised manner. Of course there will be times when he won't and then he must be reprimanded. In most cases a good growling-at will suffice, but should there *ever* be an offer of biting or kicking it must be *instantly* rewarded with a very sharp slap and a loud 'Don't you dare!' or whatever comes into your head at the time. Remember foals are like all young things —they will push their luck just to see how much mischief they can get away with, but a set of standards is essential; it makes for happier horses and certainly safer humans.

One word of warning, too, about titbits. Titbits should never be fed, under any circumstances whatsoever. They have ruined more horses than all the other things put together, encouraging biting, kicking and similar horrors.

Likewise, do not use a bucket of food for catching; it is not necessary. All our horses will come to the gate at a call and so, after they've been with us for a week or so, will most of our visiting mares, in spite of the owners' protestation that 'I'm sorry, she won't be caught without a bucket.' I wish the owners had to go out into a field containing perhaps twenty strange horses 'with a bucket'! I think they would find they could manage without it very quickly!

The other thing to avoid is too much petting and fussing. Young animals need rest and peace to grow, and if they don't get it they become irritable and bad-tempered. Particularly to be avoided are the non-horsy visitors of the 'Isn't he sweet', poke, poke, dab, dab, variety. Never take them beyond the field gate!

Feeding as the Foal Grows

During the summer months your foal will grow enormously and it is important to see that the grazing is adequate to support the mare's milk supply and her own bodily demands and also provide some grazing for the foal, who will be getting more and more of his nutrition from the solid food. Should the grazing prove insufficient due to a dry time or similar problem, the mare and foal should be brought in for an hour each day and given a hard feed. It goes without saying that ample fresh water must be available at all times.

By midsummer with its attendant flies and other discomforts, say the end of July or beginning of August, your mare and foal should be coming in for a feed every day. Bring them in to their stable about 11 o'clock in the morning and leave them until 4 pm. The box should be bedded-up with a nice soft bed to encourage the foal to lie down and sleep. Peat makes very satisfactory bedding in the summer for a box that is only occupied for a short time each day; a big bale lasts ages and there is no need to muck out completely each day, just remove the droppings and fork over. Done like this the peat dries off between each use of the box and is very economical—and the final product, if allowed to mature in a heap for six months, grows prizewinning vegetables when forked into the garden.

At this stage we do not bother to work out a separate feed for the foal but a generous feed is put up for the mare, which is then divided into two containers. The foal will soon learn to eat hard feed and before long will be clearing quite a good quantity. This whole process of bringing them in and teaching the foal to eat is part of the preparation for easy weaning, but more of that later.

Hooves and Injections

Your mare and foal should both be attended to regularly by your blacksmith. Obviously with the foal there will be little to do but it is good training for the future, that is always providing your smith is quiet with young horses. There are a few that are not, and if they are good smiths you cannot afford to fall out

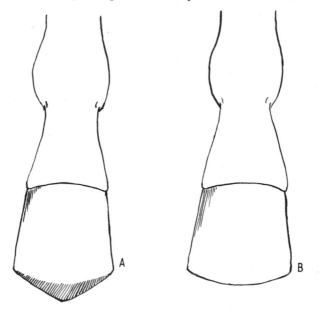

A foal's hooves tend to become pointed. The area shaded in (A) should be removed by rasping to leave the hoof shaped as in (B)

with them. So in this case it is better if you do any little bit of rasping yourself. Foals often get points on their hooves as shown in my diagram, and the only rasping required is the removal of this point, which keeps the foal walking square and stops breakages.

During the summer the foal should receive his anti-tetanus vaccination. This consists of two injections of tetanus toxoid given at an interval of not less than six weeks, followed by a booster twelve months later. These injections, coupled with biennial boosters, provide immunity against tetanus for life and should be given to every horse.

Some people vaccinate against equine flu. We do not, but take your vet's advice on this and follow it.

The other thing that must be attended to is regular worming of both mare and foal, to control red worm and also ascarids (large white worms) in the foal. These white worms look horrid but in reality are not very serious, except so far as anything that takes any share of the foal's nutrition is better not there. They are only a real danger if they are present in sufficient quantity to cause indigestion and colic. Again take your vet's advice on which preparation to use and when.

8
Weaning and the First Winter

As the days begin to shorten, the goodness goes out of the grass and it is time to start thinking about separating your foal from his dam. We normally start the weaning process when our youngsters are about five months old. You will find that your foal is really getting quite independent by then, spending long periods away from mum, grazing and going off exploring if given the chance.

If you have been following my recommended routine of bringing the pair of them in for a feed each day, the foal should be eating quite a good corn feed daily by this stage. So, particularly if the mare is in foal again, weaning should start about the end of October and be completed by the end of November, if the foal was born in May. Obviously a late foal could stay on the mare a little longer.

How Not to Wean a Foal

Weaning can be the most dangerous and debilitating time in the whole of a young horse's life. Many foals and indeed mares too are killed and injured every year in horrible accidents caused by frantic foals who have never been taught to eat or been handled in any way being weaned by being taken abruptly from their dams and shut up in a dark box with no food. A foal brought up in this way does not know that hay and oats are edible, that water is drinkable and that the bucket does not bite; and anyway the poor creature is much, much too panic-stricken and terrified to think, and will go to the most extreme lengths to get out and find its dam.

I know of a foal that was killed by jumping through a closed window; another climbed out over a five-foot wall and landed

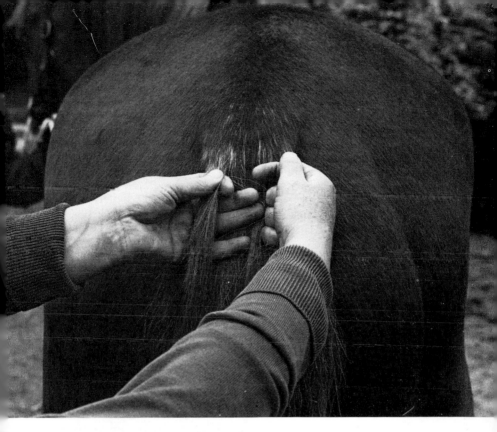

(*above*) Starting a tail plait.
The first strands of hair are
picked up as close to the top
of the dock as possible; (*left*) a
nearly-finished tail plait. The
long end is doubled up and
stitched into place

The long mane plait finished, neat and tidy, and showing off a well-shaped neck

Tobruk, a pure-bred Arab stallion, behaving sensibly in the exciting atmosphere of a historical pageant

amongst farm machinery in the next box, and so on. The list is endless, horrifying, heartbreaking and terribly expensive, both in suffering and financial terms. It costs a considerable amount of money to rear a foal to weaning age. It is unnecessary to risk the accidents and injuries that can be caused by abrupt weaning. Your mare too is likely to go dashing about and may even try to jump out to find her foal if she hears it screaming its head off in a panic.

There is no need whatsoever for weaning to be done in this heartless and dangerous manner. After all, we do not just say to our human offspring 'I'm sorry but today you are weaned' and withhold the child's bottle and milk ration. No, over a period of months we introduce solid food (oats in the case of a foal!) and gradually cut down on the bottle feeds (access to mother's milk-bar) until baby (foal) can manage without and becomes an independent being.

It is often recommended that two foals should be weaned together. In other words, when the abrupt weaning method is being used, two foals of similar ages are taken away from their dams on the same day and shut into a large loose-box which they continue to share for their first winter. I would never recommend this method to anyone, for three important reasons. Firstly, unless foals have been sharing a field and field shelter during the summer they will probably fight when first shut in together, and they may seriously injure one another by biting and kicking. This is particularly likely with well-grown colt foals and do remember in a loose-box, even a large one, there is little room for one animal to move out of the way of a kicking or biting companion, whereas out in the field escape is always possible.

Secondly, one foal is always stronger or bigger than the other, and so unless you stand and supervise every feed, the bigger and stronger one will end up eating the lion's share of both feeds. The weaker foal will become more and more deprived and is also likely to become more and more bad-tempered. This bullying may well affect the temperament of the animals concerned right through their adult lives.

The third reason that I dislike weaned foals sharing a box is that at some stage they will have to be separated into different boxes and then one is faced with what is virtually a second weaning, with all the upset and risks associated with it.

There is one other point. If two young horses become used to being always together, it can lay the foundations of future nappiness when they start work under saddle. Several years ago we had two three-year-old fillies, who had always occupied adjoining boxes, separate but with a low dividing wall which meant that they were always in contact with one another, able to touch noses, groom manes, etc. They always shared the same paddock during the day too. They were both good-looking fillies and we entered them for a quite big national show. We did our schooling, oh yes, both together in the same field. All went well, my husband leading one and myself the other. Neither he nor I gave a thought to the fact that they had in fact never been separated. They didn't share a box, did they!

Well, show day arrived and we duly entered the ring, both in the same class of course, one following the other. But here was where our embarrassment started: the judge considered one better than the other by several places. The moment the preferred filly was called into line, her friend started, to put it mildly, to make an exhibition of herself. She did everything possible, and several things that were not, in order to get to her friend. To cut a long story short, in the end the better filly was moved several places down the line to stand next to her friend. The judge, very kindly under the circumstances, gave both animals a commended card with the somewhat acid comment: 'As they are such good friends I didn't think I'd better hurt any feelings by making one better than the other.' We crept quickly out of the ring and have never, since that day, allowed any of our animals to strike up 'my best friend' relationships. They can be dangerous and very embarrassing, so please don't let weanlings share a box. They will end up either bad-tempered or impossibly nappy!

A Sensible Weaning Programme

Our method of weaning foals is a gradual one and involves separating mare and foal for an increasing length of time each day and eventually by night whilst turning them out together by day. Then, when the moment is convenient, we shut the foal in for two or three days so that mum's milk supply dries up completely. This method avoids nearly all the stress and strain with its attendant risk of injury and upset. Even more important for a small breeder, it means that mare and foal, now a weanling, can continue to share the same field and routine if necessary.

On the day you decide to start the weaning process make ready two boxes with a nice deep bed; it helps if the boxes can adjoin one another, but there should be no possibility of contact between mare and foal and it is much to be preferred if the foal cannot even see its dam. The box that is to be used for the foal *must* have a top door which closes securely and care should be taken that the window is properly guarded.

At this time of year our mares and foals only come in for just long enough to eat their food; the flies have gone, and grazing is more important to both mare and foal.

You must have assistance on this occasion. One extra person is essential, two are useful. When you go out for your mare and foal, you bring them into the stable yard as usual and the mare is led into the box that is to be used for the foal, right to the back of the box, and the foal led in after her. The mare handler leads the mare round the box and back outside. The foal handler holds the foal at the back of the box. When the third assistant, or the leader of the mare, has the door shut and the top door ready to be, the foal handler releases the foal and goes quickly out, being careful not to let the foal escape. The top door is then firmly closed and the mare put away in her box and fed.

There will be all-hell-let-loose for the next few minutes. The foal will rush about, stamp and scream and probably end up in a very bad temper indeed! Do not be tempted to go into him; leave him to get on with it and keep the top door shut. By the end of an hour or so the worst of the tantrums should have subsided and they can be turned out together again.

This is the one occasion, for reasons of human safety, on which we do not lead the foal. Take the mare out of the box and if the field gateway is in view and has unimpeded access from the door of the foal's box, take her as far as the field gate and wait whilst the foal handler lets the foal out, opening both doors at once and letting him run to mother. She should, as soon as the foal has seen her and is coming, be quickly taken inside the field. Once the foal is inside the gate and the gate shut, release the mare and get out of the way. Foalie will be in an exceedingly black mood. He has been terrified, starved and deprived, he says, and feet can fly! Actually it will take the reader longer to read the last few sentences than it will take to put them into practice!

The next day the process is repeated (a school half-term is a good moment to start if only family help is available), only the foal's feed is put in the manger as well. Over a period of a week or so the interval is increased until mare and foal are spending almost three hours parted, both being fed; by the end of this time the foal should be eating his feed quietly and possibly even lying down to rest.

This is the time of year when, particularly on a small acreage, it is convenient if the animals come in at night. When you decide to start this, you merely put mare and foal in separate boxes during the night instead of during the day. They stay out together during the day. You will find that it is not long before the foal's top door can be left open. We usually shut it when the foal first comes in and open it a crack later and gradually dispense with shutting it at all.

The foal quickly learns that he does better for feed and hay without mum's help and that he can lie and rest without the risk of being trodden on. He will see mum in the morning anyway. The mare in turn begins to like her offspring less and less, as she enjoys being free for a while; her milk supply will diminish without the foal sucking all the time and she will tend to dry off naturally, without any need to restrict her water supply or feed or risk mastitis.

By the end of a month our foals are usually quite happy to go into their own boxes away from their mothers and equally their mothers are glad to see them go. In other words they have been weaned without fuss, danger or loss of condition.

The final break can be made when it is convenient to you. Just shut the foal in for two or three days. He may get a little cross but he certainly won't pine and when he is turned out make sure it is with another young horse *who is unshod*. Make sure his dam is not in sight, or earshot too. We find that a spell of bad weather is often the deciding factor, or the mare's return to work or becoming heavily pregnant. The dam by this time will be utterly sick of him and may well warn him off her milk bar with heels and teeth, and she certainly won't fret about not seeing him.

Feeding the Foal in his First Winter

During your weanling's first winter he will need to be well fed: two good corn feeds a day, consisting of oats, flake maize, bran, soaked sugar-beet pulp and the usual additions of sliced apple and carrot, and perhaps just occasionally a little rolled barley or a linseed mash to vary the menu, and good hay to appetite. I never feed the same recipe all the time; sometimes—usually in a morning—I give just straight oats, sometimes with some flake maize added. The evening feed is usually fed damp and warm, and fairly late, about 8 to 9 pm when the youngster has eaten some hay and relaxed after spending the day turned out.

Soaked sugar-beet pulp is an excellent feed for young horses, mixed in the proportion of two-thirds beet pulp and one-third bran; it makes a very good succulent base for the admixture of oats, etc. I think everyone now knows that sugar beet pulp must never be fed dry, either as loose pulp or nuts. It must be soaked before use *If you fail to soak it you will kill your horse*. When beet pulp is dried all the water is extracted, and when the dessicated pulp comes into contact with moisture it will absorb water and increase in volume by about ten times. In other words half a normal feed-scoop of sugar-beet nuts will expand to fill a normal-size water bucket.

There is a frightening little experiment you can try. Get a small thin-walled glass tube or bottle with a screw cap, fill it with water and place one or two beet-pulp nuts in it; screw on the cap and leave to soak. The pressure caused by the expansion of the pulp is frequently sufficient to break the bottle. Imagine that happening inside a horse's stomach! So soak your beet pulp overnight. We keep a big container in which we can soak two or three days' supply at one time. Water should be standing on top of the pulp when it is fully soaked.

Once again I am not keen on made-up nuts; they are expensive and young horses often find them difficult to chew, particularly when teething, and a young horse is losing teeth at intervals throughout the first four years of his life as his milk teeth give way to his permanent incisors and molars. Often if a youngster does not seem to be doing quite as well as usual a quick look in his mouth will reveal either a loose tooth or a sore gum where a new one is erupting.

Food Additives

It is important that your weanling has access to a salt lick or else has rock salt in the manger. But expensive additions in the way of vitamin supplements are not necessary and may actually be harmful. For instance Cod Liver Oil–Vitamin D supplements can cause weak and brittle bones. Vitamin D is the substance that makes the uptake of calcium by the blood possible. This is fine so long as the vitamin is only present in normal quantities. As there is a more than adequate amount of Vitamin D in the horse's usual diet, horses rarely suffer from a shortage. If Vitamin D is then fed, a situation is created whereby the excess vitamin in the blood stream is looking for calcium to absorb. To rectify this the body mobilises calcium from the bones, which can result in brittle bones and odd bony lumps where they should not be. Likewise any other minerals or vitamins fed to excess can result in similar harmful imbalances.

Creature Comforts are Important

Your youngster should always have a good deep bed in his box as he will grow tremendously during this time and he needs to be able to lie down and get his proper rest; in fact he should spend quite long periods asleep. In the winter our weanlings go into the paddock about 11 am and come in again at about 3 pm to a well-filled hay rack. If the weather is particularly nasty they do not go out, and the mucking-out waits till the following day.

It is essential that weanlings are in at night throughout their first winter. If they are deprived of warmth, food or rest during these formative months, the growth lost will never be made up. If a young horse has to spend a winter out as a two- or three-year-old, then so be it. It is better not, but providing he is well fed he will come to no harm. A weanling will.

Routine Care and Handling

By the time your foal has reached the weanling stage, he should be quite a civilised young horse, good to catch and lead and generally well mannered. He should be brushed over regularly and have his feet picked out. Routine dosing against worms is essential and should be done every three months.

Your weanling should by now obey simple words of command: a horse is never too young to be obedient. Good manners should start at birth and grow with the horse. They should not be something that is suddenly required when the animal becomes a three-year-old. I have no patience with the 'three years old and never been touched' brigade. Surely a youngster's education should progress and widen in scope from the moment of his birth when he learns to accept humans as friends until the moment when he finds pleasure in carrying his owner on their joint ventures into the world of horsy activities.

9
Growing Up

The next two and a half years of your young horse's life should be marked by a steady progression towards maturity of both mind and body, progress which is encouraged by good food, lack of stress and proper handling.

Your yearling's second summer should be spent at grass, preferably with another youngster although this is not essential providing he has equine company of some kind. Make sure that the grazing is of good quality and adequate to support the rapid growth of a quickly maturing animal. Once the flies become troublesome in midsummer he should be brought in for a corn feed about midday so that he can eat and then rest in peace, free from torment. It is best if he can be stabled for his second winter, with a routine much as was adopted for his weanling days.

During his third summer, when he is two years old, some of his preliminary training should be done, in other words he should be mouthed and learn to accept a roller. More of that later.

His third winter is again best spent stabled at night, but if for any reason, eg shortage of space, he does have to winter out, he will not come to any harm providing he is well fed. This entails taking two corn feeds a day out to the field and waiting while it is consumed, to prevent any bullying. The containers used to hold the feeds must be heavy enough not to be easily tipped up, or else buckets or feed bowls should be placed inside old motor tyres and removed when the feed is finished. This prevents waste and ensures that your youngster gets the corn rather than the sparrows. Old porcelain sinks make ideal feed containers, being heavy with no sharp edges. Wooden boxes should never be used; if trodden on or shattered their sharp, splintered edges could cause severe injuries to legs or feet.

A plentiful supply of hay must also be fed and again if the youngster has to be outside it can either be placed in a rack fixed to the paddock fence or in a hay net, which must be tied up high and in such a way that it is impossible for an animal to become entangled in it. If you feed hay loose on the ground a great deal of it will be lost by being trodden into the mud.

It is also essential to ensure that the water supply is not frozen and that clean drinking water is always freely available. If it is freezing hard this may mean three or four trips to the field during the day to break ice and thaw pipes.

Mud Fever

In wet and muddy conditions your youngsters' legs must be constantly looked over for signs of mud fever. I personally have only rarely seen this on young horses running out. It is much more frequent on horses that are stabled and turned out by day.

Mud fever is an acutely painful skin problem, believed by many vets to be an allergic reaction to constant contact with clinging, heavy mud. The skin on the legs becomes cracked, inflamed and very uncomfortable, and if the condition is not treated promptly the skin sloughs off leaving raw patches which can be very difficult to heal. It rather resembles the effect that an east wind can have on wet hands in winter, the only difference being that humans do not have hair to mat and complicate the matter.

Mud fever can strike very suddenly and a horse suffering an acute attack can give one a nasty fright by appearing to be seriously ill. We had such an experience ourselves with a two-year-old entire colt. Obviously, being entire he was stabled at night and turned out into a paddock for two or three hours' exercise every day. He was an exuberant character who spent his time at liberty rushing about, practising handstands and pirouettes. The weather had been wet and his paddock had become rather muddy due to his energetic activities, but we had not taken any notice of that — it was early winter and quite usual for our fields to become somewhat poached, and anyway his lordship was always brushed over when he came back into his stable.

On this particular evening my husband went out to do the late-night round; we always see all our stabled animals just before we go to bed, to top up buckets, straighten rugs and check the wellbeing of each horse. He quickly returned looking worried, and asked me to call the vet urgently as he thought our colt was seriously ill with kidney trouble. I went out to see, and sure enough our pride and joy was looking very unhappy indeed, legs stretched out and swollen; he was sweating and wearing that anxious look that sick animals have.

The vet duly arrived about an hour later, not in a good humour having been dragged from his bed. He went straight into the colt's box, took one look at him, bent down and picked up a hind leg. The colt reacted by lifting his leg just about as high as it would go before snatching it away in pain—always a symptom of mud fever. 'Mud fever, that's your problem.' Light had dawned on us at the moment the vet spoke. 'I'll give you some soothing cream for his legs. There's nothing the matter with his kidneys.' Two rather embarrassed stud owners crept after the vet to take possession of a large jar of medicament.

'I'm surprised two experienced people like you didn't recognise mud fever; it would have waited until morning, you know.' This parting shot, acidly delivered as the vet's car disappeared into the night. The bill for the jar of goo and the midnight excursion came later, and was large enough to suit a night call that in the vet's eyes was unnecessary. I tell the story merely to illustrate how dramatic the effects of mud fever can be, and how quickly the acute form can occur, because that colt had shown no sign of illness earlier in the evening.

Mud fever is treated by applications of a lanolin-based cream. Udder salve is ideal, rubbed gently into the affected skin. These areas become very sore and painful and applying ointment can become nearly impossible, so we have found another easier treatment—providing there are no serious raw areas it is just as effective. Save all the old clean cooking oil from your chip pan and use it to treat mud fever and also to grease the legs of horses that have to be turned out into muddy fields for exercise. We find that if the legs of stabled horses are oiled before there is any contact with the mud, the fever does not occur. Apply it with a soft two-inch paintbrush kept for the purpose.

Cooking oil is harmless when used in this way as it is entirely of vegetable origin and is easily absorbed by dry skin. The only slight problem is that you may find your youngster becoming very popular with the local pussies in search of those lovely cooking smells!

The backs of horses wintering out must be carefully checked at regular intervals to make sure they are not becoming weather-beaten. This is a condition where the normal skin grease is produced in excessive amounts owing to over-stimulation of the skin glands by too much wet. This grease then mats the hair together, excluding air from the underlying skin which dies and sloughs off. leaving a most unsightly sore, raw, mess.

If either mud fever or weatherbeat occur it is imperative that the animal is stabled, the condition treated and housing provided for the rest of the winter. In general, horses stand dry cold much better than mud and wet. I would go so far as to say that anyone who has the misfortune to live on clinging, heavy clay soil should never attempt to winter any animal out.

Lice

Lice can cause problems during the winter, particularly in young stock wintering out. These minute bloodsuckers live mainly among the long hairs of the mane and tail. With their specially adapted mouthparts they make a little puncture in the skin and suck direct from the tiny bloodvessels in the host's skin surface. Obviously one louse is not going to have an appreciable effect, but if great numbers are present (millions rather than thousands are usual), the cumulative effect of all those minute specks of blood being lost regularly over a long period can lead to anaemia and debility. Added to this there is the irritation, which leads to scratching and rubbing, so that manes and tails often become denuded of hair. Therefore treatment is essential. All young animals, indeed all horses that have not been clipped out, should have their manes and tails regularly and thoroughly dusted with an approved louse powder. This should be worked right down to the roots of the hair with the fingertips to ensure that it is fully effective. Use it at monthly intervals all through the winter months, whether or not you can see live lice—prevention being better than cure.

During his growing-up period, regular worming is essential. Worm at least every three months and do change the product you use from time to time, as worms build up a resistance to one particular wormer after a period and some worm pills are more effective against some species of worms than others.

Your blacksmith should attend to your youngster's feet regularly and he in turn should stand quietly and pick up his feet for the smith without fuss. This however is unlikely to prove a problem if you have been picking his feet out since he was a foal. Never allow a young horse's feet to become misshapen; it takes far longer to undo the damage than for it to occur.

Anti-tetanus vaccinations must be kept up to date. Two vaccinations as a foal are followed by a further injection twelve months later. This should be followed by a biennial booster to ensure that your youngster's immunity is maintained.

At some point during this period, if your youngster is a colt he will need to be gelded. I cannot stress too strongly that this operation must be carried out by an experienced, well-recommended horse vet, whose advice on how and when he wishes to carry out the operation should be followed to the letter. I still hear, even in this day and age, of colts being treated in the most barbaric fashion and castrated by methods that one would have hoped belonged to the Dark Ages.

Tying Up

During this time he should also be taught to tie up. It is essential that all horses should learn to stand quietly when tied up but this again should be taught in a progressive fashion, care being taken that your youngster does not frighten himself by running back and breaking the rope or headcollar when he suddenly finds he is restrained by the head. A young horse should never be tied direct to the ring in the wall; this is a sure way to an accident.

We combine our tying-up lessons with grooming. We start by attaching a long rope or lead rein (about 10 feet) to his headcollar and running this through the ring on the wall, holding the spare end in the hand that is not doing the grooming. This means that whilst the colt becomes accustomed to being held by the head, there is still a little give in the rope and no danger at all of him running back and pulling against a knot. When he stands quietly without attempting to pull back, the rope can be knotted through a hay-cord loop attached to the ring in the wall, thus ensuring a fail-safe link if anything untoward occurs. Only when you are really sure that he accepts being tied by the head can the rope be tied directly on to the ring, and of course it must be tied in a quick-release safety knot. No young unbroken horse should ever be left unattended whilst tied up.

I do not approve of allowing a colt to break the hay cord tying him. Some people recommend that a youngster should be allowed to break free a few times to prevent him becoming frightened of being tied up. I cannot see the logic of this. In my opinion it is merely teaching a youngster to run back and break loose.

Early Training

If it appears that this chapter on growing up is not very detailed, it is for the good reason that during this time your youngster's life should be as uneventful as possible. He needs long days of rest, play, eating and growing in a happy, relaxed atmosphere, whilst he learns the standard of behaviour and good manners that are expected. The acceptance of this standard leads in turn to confidence in, and respect for, the human beings who handle him. I know that I keep on repeating that a young horse must be taught right from the beginning that bad behaviour will not be tolerated. It is the basis from which all further training should start.

Let us face facts. It is easy enough to ensure that a foal does as he is told, particularly if mother is already obeying the command. It will not be so easy to ensure that an unhandled three-year-old obeys you.

It is essential that any young horse is taught the simple commands, such as were listed in the previous chapter—Walk on!
Halt! Stand! Over! Trot on! Wait! etc. It is important that the
same word of command is always used for the same action.
Don't say 'Whoa!' when you usually say 'Halt!' or 'Come on!'
when you mean 'Walk on'. The tone of voice should be firm and
clear, but not over-loud. Do not be like the Englishman abroad
who imagines that everyone will understand him if he just
shouts loudly enough. A new command should be repeated in a
quiet, firm voice until it begins to be understood. I find it is
easiest to wait until the youngster is about to make the desired
action, eg to stop or to trot on, and then to use the word of
command clearly and reward with a pat for being a clever horse
and understanding.

Do not give titbits; they make for bad temper and generally
spoilt and peevish young horses; a pat and word of encouragement from a handler he trusts should be enough.

Never make any training session longer than fifteen minutes
until your young horse is at least two, and even then only increase
the length of the lesson if you are doing something definite, such as
going out for a walk. A youngster's attention and enthusiasm
will only last for about this length of time, then he will start to
get bored with negative results in the way of progress made.

Should your youngster be sinful he must be reprimanded at
once and left in no doubt at all that whatever it was he did was
unacceptable. I know I have said this elsewhere, but any
attempt to bite or kick must be instantly rewarded with a
sharp slap. Colts frequently go through a stage when they nip.
This must be firmly discouraged by a quick tap on the nose, and
should there be any indication that the nip was other than
playful, punishment is called for.

If your youngster does misbehave, and they all do at some
time, it is no good reprimanding him half an hour later; he
won't remember what it was that displeased you. He will simply
wonder why you've suddenly lost your temper with him and
his trust and confidence in you will be severely shaken.

The most unforgiveable thing I ever heard was told me by a
woman whose two-year-old colt bit her hard when she was

leading him in from the field. Now, instead of reacting as she should have done and giving his head a good shake (catch hold of the back of the headcollar and shake hard until his eyes rattle and growl at him—it works wonders!), she—as she put it—kept her temper and went indoors and fetched a stick and gave him a good beating. What a terrible admission; what a jolt that youngster's confidence suffered! In fact the horse in question was never reliable to handle after that incident.

I believe that if an animal gets stroppy with you, you are entitled to get stroppy back, but never, ever lose your temper and never, ever hit an animal round the head or legs. Never under any circumstances, take a stick into an animal's stable to beat it with—and never, ever, tease a horse. If you can't manage any horse, stallions included, without recourse to weapons, you are not a fit person to be in charge of one. I do not of course mean that you should not carry a cane when riding or showing in hand; that is a different matter. In these cases a stick is an aid, not a weapon. Another point: try to avoid having a disagreement with a young horse unless you have the time, patience, cunning and assistance to make sure the victory is yours—and yours without loss of temper!

A three-year-old of any size, horse or pony, is very much stronger than his human handlers. If any animal is frightened by unaccustomed or rough handling he will obey his innate instincts and use all the means he has at his disposal to ensure that his captors (and this is how he must regard human beings if he has never learned to trust and respect them) cannot torment him further. In the case of a horse the result is likely to be that the youngster learns to bite, kick, rear and possibly even strike out with his forefeet, and dangerous though these things are, the real tragedy is that the young horse who behaves like this comes to realise that he is physically stronger than his handlers. Once he has learned that he can get his own way by sheer brute force, all further training becomes a risk and a battle. The whole premise of teaching a horse to carry someone on his back and do his rider's bidding is based on the precept that the horse must never be allowed to realise how strong he is. We maintain our control over any horse only by superior intelligence; none of us can do it by strength. We must therefore use our intelligence to ensure that our youngster does as he is told. So often force and blows are used by ignorant people in an attempt to make a youngster learn in a month what he should have absorbed gradually over three years. Rather than teach a youngster anything, the bash-and-crash brigade invite rebellion. We prefer that our youngsters' education should be progressive and based on mutual trust and respect.

Handling

Two points should be mentioned here. Firstly, never attempt to lead any horse from the headcollar alone without a rope. The slightest flick of the horse's head and you will find it impossible to keep your hold on the headcollar—and hey presto! your youngster is careering about loose and before you can count ten an accident has occurred. Always lead an animal by a rope or lead rein. There may be occasions when you sense that your passage to the field may not be going to be a smooth one. Perhaps your colt may be over-fresh after being in for two or three days owing to bad weather. In this case it is a good idea not to fasten the lead rein but to loop it through the headcollar and catch hold of both ends so that there is nothing trailing if he does get loose; quick release is also possible.

An Emergency Halter

There is bound to be at least one occasion when a young horse gets loose without a headcollar on, and it may be just the time that for some reason no halter is immediately to hand. Many years ago our vet showed me how to make a very effective emergency halter from a lead rein, a rope or even at a push a length of hay cord or stout string. The piece of rope, etc, should be at least 6 feet long and have a loop securely knotted into one end (use the handle end of a lead rein).

Everyone knows that it is often possible to grab a loose horse either by the mane or round the neck, but it is quite a different matter to actually put anything over the miscreant's nose and

Mouthing a 2-year-old. The mouthing bit is tied on to the headcollar

First steps forward with a rider. The 3-year-old in the picture had been backed three days previously

A young horse learning to tackle small obstacles. Note the hand in the horse's mane

A young horse crossing water happily. Her sire is acting as escort

then fasten the headpiece. All too often the loose horse remains loose and there is a risk of the would-be catcher getting kicked or trampled on, particularly if the animal is excited or frightened and not thinking straight. This emergency halter has the great advantage that the first move is to put the rope over the animal's neck so that the loop hangs down below the neck; the rope should be placed fairly close behind the ears, as in the drawing. This can be done with an animal that is moving, even at a trot. A loop of rope is then pulled through the hanging loop A, and this loop is passed over the animal's nose. The rope end must not be pulled through loop A.

This makeshift halter gives considerable control as it acts directly on the horse's nose. Care should be taken, however, that the pulled-through loop does not pull tight, as this may cause resistance in a horse unused to pressure on the nose.

There is another advantage, too. Apart from being quick and simple to make and effective, the halter is quick and easy to remove: quite a point if, for instance, the horse is being returned to a field where there are other excited horses milling around threatening to trample one under foot. The nose loop is simply slipped off and pulled back through loop A, allowing the rope to slide away from the animal's neck.

Practise quietly a few times until you can make this emergency halter without thinking how. Our experience is that once learned this makeshift restraint will be found useful in many non-emergency situations. It is equally efficient for dealing with errant cows, donkeys, goats and, at a push, for a single cabbage-eating sheep. I've not tried it on a pig!

Whilst I am on the subject of quick-release and emergency halters, it surely goes without saying that every horse should be turned to face his handler before being released in his field. This means that on being set free he has to turn round before he gallops off and that in turn gives the leader time to step back and out of the way of the bucks and kicks that all horses, particularly youngsters, indulge in when they are feeling fit and well.

The other thing to remember is that if more than one horse is being turned out at a time, no one releases their animal until everyone is ready and has said so. To do anything else is inviting loose horses with lead reins trailing and kicked or injured leaders.

When leading a horse into a loosebox always lead the animal right into the back of the box and turn it round to face the door before releasing it. Then you do not have to push past the sharp end in order to get out, and apart from that you are always between the horse and the door, so that unintentional escapes are avoided. Our horses all automatically turn to face the door and stand and wait whilst their headcollar is removed. I consider it to be gross bad manners if an animal bulldozes its way to its hayrack and starts munching before being released.

It is also a good habit to give the loosebox door a pull towards you after latching it on your way out. It is surprising how often the bolt fails to connect and if the door opens as you pull, the matter can be rectified with no harm done, rather than finding a loose horse in the yard at midnight.

One last word on the subject of loose horses. We have all encountered a stray horse wandering along a country lane or, even more frightening, on a main road. Of course one's first reaction is to stop and try to catch the wanderer, emergency halters uppermost in mind. Did you know that if you take charge of the animal—by trying to drive it into a field, for instance—and while you are so doing it gallops off and causes an accident, you could be held liable for any damage it causes? If by some mischance someone was injured, substantial sums could be involved. It seems unfair, but as no one could stand by and see a horse cause an accident if it could be caught and led to safety, it seems necessary that everyone who has dealings with horses should have adequate third-party insurance.

Many people advocate lungeing a young horse, either for exercise or as part of his education. No dressage rider ever asks a

Loop A (at least 3" long)

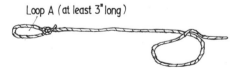

An emergency halter. Take a 6ft length of rope, lead rein, hay cord or whatever is handy and make loop A

Place the rope over the horse's neck, as near to the poll as possible, and so that loop A is level with the bottom of the neck

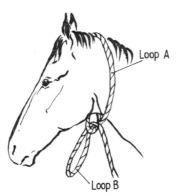

Pull another loop, B, through loop A, as shown. Do *not* pull the rope end through loop A

Place loop B over the horse's nose. (To remove, slip loop B from the nose and pull—the rope will fall clear)

young, unfit and inexperienced horse to work on a close circle, as the strain on immature and weak joints and tendons is far too great. Why then should we subject an even more immature animal who has not even finished growing to the strains that a riding horse in the early stages cannot absorb? I believe that no horse under the age of three years should ever be lunged. Much better to take them out for a walk, but more of that later.

The other thing I have seen unthinking people do is to chase youngsters up with an umbrella or polythene bag in order to get them to show off their paces. I ask one question: what happens on the day when your colt, just backed and inexperienced, meets someone carrying a brolly, or sees a polythene bag in the hedge—and farmers do not always dispose of them very carefully? The mind boggles. For my part I will do without the extravagant show and preserve my safety and my seat in the saddle.

10
Preparation for Further Training

The education of a young horse must progress slowly over a period of many months. A great many useful animals are totally ruined by being badly and carelessly broken in, in far too short a time. I become angry when I hear people who consider themselves knowledgeable declare that all the young horses they break will jump 4 feet and do a full pass six weeks after having been sat on for the first time!

A youngster must be given time for his muscles to adapt to the strain of carrying a rider. He also has to learn to balance the weight of that rider. It takes at least twelve months for physical fitness, balance, ability and stamina to develop to the stage where a horse is ready to start specialised training for jumping, dressage or eventing.

Many of the evasions, vices and general bad habits, such as incorrect head carriage, hollow stiff backs, inactive quarters, bucking, bit evasions, jogging and generally poor paces have their origins in muscle fatigue; they often result after a young horse in a poor state of physical fitness (too fat being worse than slightly lean) has been asked to do work for which his muscles are not yet capable. So-called suppling exercises on a small circle are a case in point: the effect is obviously worse if the rider is not very good and rides in a heavy, lumpy fashion with unsteady, rough hands.

Getting someone on to the back of a young horse is the easiest part of the training process. There is much to be done before and after the actual backing.

We always start by mouthing our young horses and walking them out in hand. This is important for two reasons: firstly, it teaches a colt to accept the bit and also to use contact with the bit as an aid to balancing himself; this obviously lays the foundation for a good mouth in later years. Secondly, walking a youngster in hand gets him used to traffic and the sights and sounds of his environment with someone beside him to give confidence and encourage him. It is a fact that none of the young horses that we have broken has ever been any trouble in traffic when first backed, or afterwards as mature animals.

Mouthing

We always mouth our youngsters as two-year-olds during their third summer. By this time they are strong and well grown and the extra control provided by a bit is necessary, particularly if they are going to be walked out on the road, or shown. This mouthing process normally takes about six weeks and involves a progression from learning to open the mouth and having a bit put in to learning to obey signals given by means of the reins.

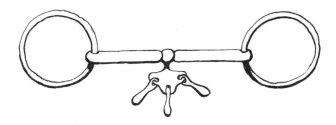

Mouthing bit with keys

It is a good idea to teach your youngster to open his mouth before the bit is introduced, then when you put the bit into his mouth for the first time you don't have to get it past clenched teeth with the attendant risk of banging his gums and causing bruising. You will need help for the first time or two that the bit is put in. You need a stout leather headcollar that fits the animal, a mouthing bit with keys, and two pieces of stout string or straw cord about 12 inches long. Put the headcollar on and adjust it so that it fits fairly tightly; do not attach the bit. Have a lead rein on the headcollar which your assistant holds to keep the animal still. The assistant stands on the offside of the horse. The bit, which can usefully be smeared with a little treacle, should have a piece of string or straw cord securely looped through each ring. Taking the bit in your left hand, open the horse's mouth with the right hand and being careful not to bang the teeth, pop the bit in. When it is in place, both pieces of string are fastened to the Ds on the side of the headcollar, and tied securely by your assistant on the offside. (Bit straps are the proper thing but not everyone has them.) Make sure that the bit is in the proper place in the horse's mouth, adjust the headcollar a hole or so if necessary, release your horse (leaving him in the loosebox of course) and go away and leave him to cogitate upon life and the strange ways of humans for an hour.

The process is then reversed; the strings are untied, the horse's mouth opened and the bit removed. I make the point that the mouth should be opened; most youngsters are only too glad to drop the bit, but you get the odd one who clenches his teeth and refuses to let go unless you open his mouth. It is important at this stage that the assistant again puts the lead rein on the headcollar and prevents the colt rushing off round the box or, possibly worse, the handler grabbing the bit for restraint. This routine is repeated for the next two or three days and not until the colt opens his mouth confidently and willingly is the bit secured to a headpiece.

Never try to get a bit in a youngster's mouth on your own. You must have help to steady the horse and tie the bit to the headcollar on the off side. Do not tie the bit to the off side D of the headcollar and try to pull it up into the horse's mouth that way. It only needs the horse to fling up his head and get loose and you have a terrified youngster rushing round the box with a clattering metal object swinging loose round his head—and then at best a permanently head-shy horse.

Once you are certain that the colt will open his mouth and accept the bit, it can be secured to a headpiece which can be put on over the headcollar. No browband or noseband is necessary, just the headpiece of an old bridle which does mean you have a throatlash to provide anchorage. The lead rein is still attached to the headcollar every time the bit is taken in or out of the colt's mouth.

Going out for Walks

Once this stage has been reached it is a good idea, if it is at all possible, to start taking your youngster for walks. He is still led from the headcollar but with the bit in his mouth. A suitable first walk is to the end of the farm drive, say about 500 yards and the same distance to return.

Insist that he walks out freely beside you, straight and on a loose rein. It is also a good idea to vary the side he is led from. I always take our youngsters to the end of our lane and let them watch the traffic for five minutes or so, making sure that I have someone with me to warn of the advent of anything particularly alarming so that I may beat a prudent retreat if necessary.

After about a week reins may be put on the bit, and over the period of about a fortnight your colt is accustomed to being led from the bit. To start with the reins are just held in the left hand and the lead rein is still used by the right hand for control, but gradually the reins can be taken up by the right hand until eventually your youngster is being led from and responding to signals from the bit.

The lead rein is still kept attached to the headcollar which is worn under the bridle right through the whole mouthing process. The lead rein can be used if firmer methods of control are needed, for example a good tweak on the headcollar by means of the lead rein is better than a jab in the mouth if the brakes suffer from fade when a polythene dragon appears in the hedge. During this time the daily walks are extended, and indeed, providing traffic conditions will allow, your youngster will enjoy going out and about around the lanes and investigating strange sights and sounds.

I always take a helper with me to warn traffic of my approach. It is unfair to other road users to go out alone with an unpredictable youngster and expect instant co-operation from them when they meet you suddenly on a corner. Most drivers are most considerate and will pull up while you lead your youngster past their vehicle provided that they are asked politely and in good time by an escort on foot who can explain that you are leading a young horse. Better only to walk twice a week, perhaps once mid-week and again at the weekend, with a helper than to risk going out alone and causing an accident.

One more point here. With traffic approaching from behind, until you know your youngster is reliable, ask the car to stop, turn round and lead the colt back past the car, thank the driver and then proceed as before. It is far safer to lead a youngster past a stationary vehicle and also better if the vehicle and horse are pointing in opposite directions, because then if the colt should jump forward he will jump out of trouble, not into it, as would be the case if horse and vehicle were travelling in the same direction.

If you meet a real horror, lead your charge into the nearest field and stay there with the gate closed until the combine, steamroller, tipper-truck or whatever has passed. Avoid at all costs getting your colt trapped in a gateway or similar place with nowhere to go and a large vehicle approaching. Much better if needs be to turn round and trot ahead until a suitable field or lane can be found to disappear into or down. Try to get your colt turned to face the horror before it passes you.

It is often a good idea to go for your walks during the lunch hour. It is my experience that traffic is at its lightest then. I should not need to say this but do thank all drivers who stop for you or make your life easier in any way. At no stage in this process can you go out unaccompanied. As your youngster's education progresses, your assistant can ride a quiet and reliable pony, which gets your colt used to company and also provides a lead should you require one.

I fully realise that many people have their establishments near to main roads and heavy traffic. Obviously it would not be safe or sensible to walk a young horse in such conditions. Equally obviously, where you live you ride, and unless all your riding time is to be spent in a covered school or jumping paddock, your young horse has to face the traffic at some time and I would go to any lengths to find a farmer with three or four fields adjoining a busy main road and ask his permission to walk my colt in those fields, with the protection of the fence between myself and the traffic. It is worth boxing your colt there if necessary. It is essential that all horses are as safe in traffic as it is possible to make them by sympathetic training, preferably before there is a rider at risk on their backs.

Maturing

With all the walking exercise you will find three things happening. The first is that your youngster will start to alter in shape as his muscles develop and you will get some inkling of what the final horse is likely to resemble. This is particularly true if your colt has been carefully 'mouthed' and gradually comes to accept the bit and drop his head into it. It is for this reason that I always use reins on the bit and not a lead rein. Aids much more closely resembling the rein aids that will be used when he is eventually ridden can be taught, a light and sensitive contact being maintained at all times.

The youngster's paces should also start to mature. At all times insist on free forward movement with a long free stride, both when walking and trotting. Remember always to use the same clear words of command, eg Walk on! Trot on! Wa-a-alk! —said slowly to pull up from a trot. All transitions must be clean and precise and forward-going. It is little trouble to insist on this whilst walking out and it does lay the foundations for ridden work.

The second thing that may happen is that your colt becomes footsore. If this happens, rather than curtail his work ask your blacksmith to fit a set of light shoes which should last for about six to eight weeks, if removed once. The fitting of a set of light shoes should pose no problem providing your youngster is accustomed to having his feet trimmed and handled, and it is much better than allowing him to associate going out with painful feet.

Be careful that your horse does not become alarmed the first time he hears the sound of shod hooves on concrete. Walk around the stable yard a time or two before going out for the first time after his shoes have been fitted. He will soon realise that he is only listening to the sound of his own footfalls and that there is no one behind him. It is also a wise precaution to fit a set of brushing boots just in case he knocks himself while becoming accustomed to walking and trotting with shod feet. Make sure that they are fitted properly, though, and that there is nothing to slip or break loose and frighten him.

The third thing that will happen is that your colt will start to get fit, and may even become a bit of a handful. At this stage it is a good idea to introduce a roller. This must be done, like bitting, gradually, and no matter how confident and trusting your colt, there may be fun and games the first time he feels something tight-buckled around his middle. Again have a helper, introduce the roller slowly and do not girth it up tightly for several days.

A good routine is to take him for his walk, bring him home, let your helper hold him by the lead rein on the headcollar and show him the roller. Place it over his back and just hold the ends together, girthing up taking several days longer.

Once the girth is fully tightened your colt should be asked to walk forward a few strides and this is the moment when, no matter how quiet he has been, there just may be an explosion. For this reason it is essential that the introduction of the roller is done in an enclosed space such as a schooling paddock or covered school. Roads and lanes are not the place to try to cope with an aerobatically minded youngster. He must have accepted the roller completely before he is walked out in it. It is better to take him for his walk and then spend five minutes on roller lessons each day, better to achieve five or six strides without a rodeo, than attempt 200 yards and have him loose after 25 yards.

As with all 'new' pieces of tack, the roller must be introduced slowly and carefully with a little progress being made each day. Remember that any untoward incident will take many days to be forgotten. 'Make haste slowly' is an excellent motto when dealing with young horses.

Remember too that the roller only needs to be tight enough to stay in its correct position; it must not be so tight that it resembles a tourniquet round the animal's chest—that is inviting a colt to buck. It should be tightened gradually and gently when you put it on; you do not pull it up to its final hole in one quick movement. That again invites reaction. The correct final fitting should allow room for three fingers to be inserted between the girth and the rib cage.

Once he is accustomed to the roller it is a great aid to control when walking out if a pair of side reins are fitted. These should be fitted so that they do not act unless the colt's head comes well out of the proper place, eg down between his knees or straight out in front. In short they are merely there to prevent him straightening his neck and getting a rise on you; better this than heaving on the mouth of a youngster in a desperate attempt to pull him up from a trot.

By the end of two months, before the flies, heat and holidays arrive, your colt should be mouthed, and wearing a roller or even an old saddle without fuss. His paces should be good and freely forward-going. He should know and obey simple rein aids and all voice commands, and be quiet and sensible in traffic and fairly confident about the various sights and sounds he is likely to meet in the environs of his home. This is the moment to turn him away until he is three, but if his two-year-old education has been patiently and thoroughly carried out, an invaluable foundation for his future career under saddle has been laid.

By the time your youngster gets to his third summer he should be a well-grown, well-mannered, sensible and confident young horse, who leads well, obeys simple words of command, wears a roller and possibly other items of tack without complaint and is properly mouthed and obedient to the reins: in other words, fully ready to progress to further education.

I I
Showing

At some stage during your youngster's progression from foal to riding horse, a local sage is sure to say, 'You know that mare and foal, yearling or whatever, of yours is good enough to show. There was nothing at Blankshire show nearly as good.' The evil seed has been sown, the idea nags away. Visions of large rosettes and silver cups become part of your dreams, and before you realise what you have done the entry form is filled in and sent away. So be it, but let me offer a word of warning. It is a very good thing for every young horse to attend a few shows during his first three years. It accustoms him to being plaited and trimmed, being boxed and travelling to strange places, meeting unknown horses and people in unfamiliar places. But shows are very tiring and no young horse should be asked to compete in more than two shows a year if he is not to become tired, stale and sour. It is a fact that very few horses who have been shown extensively in-hand as youngsters ever progress to make good riding horses.

Having said that, let us return to the procedure for making your show entry. Study the schedule carefully and enter your horse in the most suitable of the classes for which he is eligible. This is not always as straightforward as it seems; at nearly every show the unfortunate judges and stewards have to contend with animals that have been incorrectly or unsuitably entered—colts in filly classes and vice versa, ponies in hunter classes, hunters in hack classes and so on: the possibilities seem to be endless! Actually the mistake most usually made by the inexperienced breeder is to enter his exhibit in a class for which it is eligible but totally unsuited. Examples which come to mind are the hunter and small hunter classes. The schedule usually lists 'Yearling, two- or three-year-old filly, colt or gelding likely to make a hunter' and, for the small hunter, 'Yearling, two- or three-year-old colt, gelding or filly likely to make a small hunter or hack not to exceed 15.2hh at maturity'. Think carefully about which class is the more suitable for your youngster and unless you are expecting him to mature at well over 16 hands (ie to be over 15 hands as a yearling), settle for the small hunter class. Most youngsters by a thoroughbred or Arab sire out of either 'cob' or native-pony-type dams mature at between 15.1 and 15.3 hands. If they are then shown against show-hunter young stock they will look totally out of place and insignificant. Add to this the fact that many show animals are born very early in the year and are thus six months ahead of their rivals in maturity, and it must be obvious that if a general-purpose family horse is entered in an unsuitable class it is inevitable that he will be overlooked. Remember too that most horses are at least 2 inches smaller than their fond owners imagine.

The other pitfall that sometimes seems to trap the unthinking is the definition of 'of pony type' and 'of riding horse type'. Pony type must show definite pony characteristics; so if you enter a 15 hands small hunter in such a class you are likely to be ignored. One other thing: if a class is for registered 'anythings', be it mountain and moorland ponies or part-bred Arabs, it means what it says—and the animal must be fully registered with the appropriate breed society.

Having made your entry, there are several things to be thought about. Your youngster must be taught how to trot out and how to stand to show himself to the best advantage. You must consider what tack he is to be shown in and if he needs mouthing before the show. The blacksmith must be asked to trim his feet and any last vestige of winter coat must be removed if the show is an early one. Your exhibit should be stabled at night for at least a week before the show and given two corn feeds per day. The extra energy coming from the hard feed will be needed to sustain your animal during a very tiring day. Do not even consider entering any animal for a show if it is not in good condition. It is simply a waste of time and money and will merely prove an embarrassment to you. Everyone's exhibits always look better at home, and so if you have any doubts about the preparedness of your horse, either in condition or training,

Carefully placed plaits can improve the look of a horse's neck. (Above) On a fat neck, small plaits are placed below the crest. (Below) On a thin neck, larger plaits are arranged along and on top of the crest

stay at home, even if you have already entered. It is much better to forfeit your entry fee than have everyone saying 'Why ever did they bring that thing here?'

In your enthusiasm to get your animal ready to show, do not overfeed and end up with a youngster so grossly overweight that the fat wobbles when he trots. As a judge I always penalise severely any exhibit that is too fat for me to see its shape!

Two other things need to be considered in good time. The first is transport. If you are hiring a box, be sure to make the booking as soon as you have made your entry, stating clearly what animals you expect to be taking and what partitioning you will require. A mare and foal need sufficient room to travel loose in safety. The second thing is plaiting. This should be practised a time or two to get your horse used to the process, and to enable you to see where the plaits can be placed to best advantage. Careful placing of suitable-size plaits can do much to improve the shape of an animal's neck. If you find that you really cannot make a tidy job of plaiting a mane, try to find a friend whom you can persuade to do it—nothing looks worse than plaits growing long straggles, or falling out at the crucial moment.

Now to think about things in more detail. It is most important to teach your show entry what is required of him and how to do it properly, making sure that you also know what you yourself will be expected to do on show day. If you are a complete novice at the showing game, visit a show or two as a spectator and watch and learn. Practise leading your entry until he walks and trots out freely beside you on a loose rein. He must also be taught to stand properly, still and square on all four feet, with his head and neck in an attractive position. No judge can get a proper look at an animal that will not stand still and when he does momentarily come to rest adopts a posture that suggests he is first cousin to a camel.

Trotting a horse out in hand is something that also needs practice. It is embarrassing in the extreme to have a horse that drags along behind you as if his feet were set in hardening concrete, or worse still refuses to trot at all: this may well lead to a helpful ring steward chasing him up with a programme, umbrella or mac, and your forward progress may then be faster than you intended and accompanied by a certain loss of dignity! Be careful not to trot too fast. Some inexperienced exhibitors do this in the hope that their horse will produce a brilliant extended trot. This could happen, but what is more likely is that your horse will either end up cantering or lose his balance and start going all on his forehand in such a way that he appears to move very wide behind—most ugly. Aim for a medium-paced balanced trot with the hind legs providing the impulsion. Once your youngster is moving in this way, the chances are that if you extend your stride he will extend his stride also (don't forget you should always run in step with your horse). This also entails the leader being fit enough to run fairly fast beside the horse for at least 200 yards.

Having made these general remarks let us go through the complete show-ring routine. The collecting-ring steward will give the order to lead on into the ring, usually and correctly in catalogue order. If there is no set order try to follow one of the less good entries in the class; avoid following an animal that is a multiple championship winner. I'm not saying that your exhibit isn't far superior but . . .! Walk into the ring in a confident manner and lead round in a clockwise direction. Try to give yourself room to get your horse walking freely, avoid getting crowded and keep well out. Judges can suffer from intense claustrophobia when upwards of thirty young horses keep closing in. Keep at least two lengths away from the horse in front; strange horses can and do kick and it is better to be safe than sorry. Some judges ask the class to trot on along one side of the ring as individuals; some do not, and start calling competitors into line after three or four circuits of the ring. Keep your eye on the ring steward and wait to be called in. When your call comes, acknowledge that you have noticed the steward. Gentlemen raise their hat, ladies their cane, and go quickly into line. Your exhibit must stand properly as he has already been taught; picking a little handful of grass or having a nut or something for him to seek in your hand will ensure that he pricks his ears and drops his head, thus ensuring an attractive outline to his head and neck. You will be called out in turn for the judge, when again your horse must be stood out correctly. Keep him standing to attention whilst the judge walks all the way round him and has a good look at all the component parts.

The judge will then ask you to walk your horse away from him and then to trot back past him. The walk away should be straight, free-going and on a fairly loose rein; be careful not to pull your youngster's head towards you, as the judge is looking for straightness of action and a turned head can affect an animal's action so badly that it makes him appear lame. Do not go on walking for ever; about fifty paces is ample. Pull up and turn your horse, pushing him round on the inside of you. If you pull the horse round you, firstly he will become unbalanced and on his forehand, and secondly there is a distinct likelihood of your getting trodden on or knocked over. Complete your turn walk two or three strides and then give the order to trot. Doing it this way will ensure that your youngster starts his trot straight and balanced. Do use a word of command, not the ubiquitous clicks and clucks. 'Trot' or 'Trot on' sounds more professional and is clearly understood. I have been to shows where the in-hand classes sounded like a day trip for battery chickens, all cluck and ruffled feathers. Strike off into a run beside your charge, making sure that you run in step with him. Trot past the judge and on around the rest of the line of horses, keeping going until you reach the other side of the ring; it is surprising how many rosettes have been won by giving a judge a quick view of a nicely balanced youngster really stepping out down the far side of the ring. Pull up and return quietly to your place.

On the subject of trotting horses out for the judge, I must tell a little story. Several years ago I was at a major national show showing a two-year-old colt. We had done our show and returned to the line, and I watched with some amusement as another competitor went down to start her trot. It had obviously been impressed on her that a really good show at the trot was essential. She pulled her colt up, turned him round and started to run back towards the judge, extending the lunge rein that she was using to lead her colt as she ran. There was only one thing wrong: her colt didn't move. He stood rooted to the spot looking with utter astonishment at his handler as she disappeared into the distance. This unfortunate lady soon came to the end of the lunge rein, was pulled up short, tripped over her own feet and fell flat on her face in the grass, whereupon her fiery beastie put his head down and started to graze. To this day I can still see the amazed expression on that colt's face. I expected to hear him ask 'Mother, what are you doing down there?' So do your homework beforehand, rehearse your show and avoid such untoward incidents! You may well find that the strange atmosphere, sights and sounds of the show ring have sufficient effect on even a well-trained youngster to ensure that you have your hands more than full.

After the line-up the judge will ask those in the running for final placings to walk round again. All the class may be included, or else only a dozen or so, with the remainder being told to leave the ring. At this stage it is important to watch the steward carefully and obey his instructions at once. Accept whatever placing you receive with courtesy and good grace, regardless of whether you finish at top or bottom.

If you win or are placed, be sure that you take part in any judging for special prizes or championships for which your win may have made you eligible. Usually only first or second prize-winners are required for championships. At some shows there is a parade of prizewinners and if required you must parade your animal. These parades can be either very trying or good experience for a young horse, depending on your frame of mind at the time. The ring not infrequently becomes filled with everything from beef bulls to donkeys and even pigs and tractors! Amongst horrors I have personally encountered have been the champion pig of the show, a huge large white boar escorted by two white-coated attendants, a brewer's dray drawn by six magnificent black shires, all jingle and rattle and quite overwhelming for a yearling filly, a loose Hereford heifer, and army dispatch-riders entering the ring as the tail end of the parade left by another exit. As I say parades can help to broaden a young horse's outlook very considerably!

One final cautionary word. If you do go home laden with cups and rosettes, don't get carried away and enter for every show within miles. Better to be content with two championships a year and retain a happy contented horse!

There are several other aspects of the showing game that need careful planning and rehearsing before the day of the actual show. Do accustom your youngster to someone strange walking round him, someone who may, on show day, be wearing a large hat, sunglasses or bowler, etc. It is not helpful if your horse becomes terrified of the judge! Make sure too that he is used to someone walking behind him: it would be an unforgiveable sin if he kicked the judge.

If you are showing a mare and foal, see that the leaders of both of them know what to do when the mare is trotted up for the judge—is the foal to be trotted back after the mare, or taken to the top of the ring to wait for the mare? Practise and see which method works best for you, and stick to it on show day.

Think also about what tack your horse is to wear on his head. On a good head, nothing looks better than a fine dark in-hand bridle. Personally I dislike coloured or brassy browbands, but this is a matter of individual choice. If your horse has a rather long head, it can be made to look shorter by wearing a wider noseband and possibly a white browband. A plain head can be disguised by the use of a heavier bridle with plenty of brass buckles and a bit with large rings. See that the bridle is properly cleaned several times before the show so that it is easy to adjust and nicely supple; all brasswork should be polished till it sparkles, and when the bridle is given its final clean before the show it should be put together so that it fits without further adjustment at the show. Make a note of which hole everything goes in when the bridle is on the youngster's head and fitted properly.

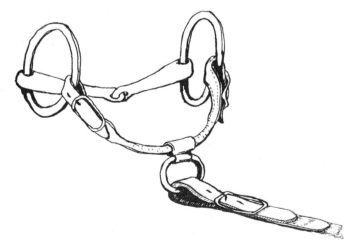

Method of attaching coupling and lead rein to the bit of an in-hand bridle. Note: no chains or clips to undo, everything buckled and secure

A 10- or 12-foot lead rein is quite adequate for leading any horse. If you use a lunge rein or similar you may trip yourself up in the spare footage. A lead rein with a buckle fastening is far safer than one of those clanking monstrosities that are all brass chains and clips. Chains can cause horribly painful injuries if they happen to be pulled rapidly through your fingers and I have actually seen a nasty accident in which a yearling filly lost two front teeth and tore her mouth most terribly when a lead rein ending in a brass chain, the kind which divides into two chains each with a clip, got over her lower jaw when she tried playfully to chew it. This cannot happen with a lead rein that buckles on to the back of the headcollar or on to the special buckled couplings on the bit.

Another thing about clips. They undo, and a loose horse in the show ring is both dangerous and embarrassing. Don't fall into the trap of whitening the lead rein; just scrub it and leave it off-white. It may not look quite so smart but it will stop you resembling a decorator's mate if it happens to rain and the whitening gets wet and runs.

If you have any doubts whatsoever about controlling anything other than a foal in just a show headcollar, the animal must be mouthed about six weeks before the show (see previous chapter), shown with a bit in his mouth and led from the bit. The lead rein should be attached by suitable couplings to both sides of the bit, not fastened to the off-side ring and pulled through to the near side. This has a very severe curb effect and should never be used on any young horse. If you are proposing showing an entire colt of the bigger native breeds, it is customary—and indeed necessary—to show him in side reins and a roller, and again this tack should be introduced well before the show.

Think also about what clothes you are going to wear. Gentlemen should wear either a suit or hacking jacket and trousers, with a collar and tie; no open-neck shirts, please. The outfit should be completed by either a bowler or soft tweed hat, gloves, a cane (plain leather-covered) and sensible shoes in which you can run. Lace-up shoes with non-slip soles are really the only choice. Ladies must never wear open-toed or strappy sandals—it is very dangerous, as such shoes offer no protection from a misplaced hoof; having a bare foot trodden on is not only exceedingly painful but could result in broken, crushed bones that could cripple you for life. The height of stupidity was a bare-foot female taking a stallion into the ring at a major breed show a few years back. The stewards quite correctly ordered her to leave the ring. A lady needs also either a dark showing jacket or a hacking jacket, worn with a neat collar and tie and plain well-cut slacks without flared bottoms. If you don't do a lot of showing a well-cut trouser suit looks smart and is quite acceptable. Again a hat of a suitable type should complete the outfit, with of course gloves and cane as before. Avoid the headscarves so often affected by horsy females—they look sloppy and untidy and, what is worse, because they are tied tightly over the ears the wearer often cannot hear what the judge or steward is saying. Hair of course should be tidy and tied back or otherwise incarcerated so that it cannot blow about.

Plaiting your horse's mane and tail should be practised a time or two before the great day, as already said. Most people can make a more or less tidy job of plaiting a mane, but the tail seems to cause problems to the most unexpected folk. For myself, I find this particular task far easier than fiddling about getting mane plaits even, and then having the needle unthread just as you have persuaded your yearling to hold his head down so that you can stitch in the last plait! Give me a tail any time. As I do not like pulled tails—the pulling process can make an animal's dock sore, setting up an irritation which can lead to all sorts of problems, and a pulled tail also looks a terrible mess when it starts to grow out—I plait tails. My photographs will show you how to do it. Practise until you can get the tail looking like the final illustration. On show day, plait your exhibit's tail and put on a tail bandage. Do remember, though, that when you get to the show the bandage must be unwound, not pulled off, or your plait may not look quite so good.

There is one other problem that owners of occasionally shown youngsters may find bothersome—what to do with a long luxuriant mane if you are only intending to exhibit at one local show? It is hardly worth pulling your horse's mane to length;

Plaiting a long mane: (1) take three 1-inch strands of mane and start an ordinary plait; (2) continue this plait for three or four twists; (3) add in further 1-inch sections of mane and continue to plait; (4) pull the plait to lie along the length of the neck; (5) continue plaiting until there is no more long hair to plait in. Finish the plait in the usual way and stitch the end securely; (6) any short hair can be made into a separate small plait

the process itself will have to be drastic, which again may result in soreness and irritation of mane, horse and handler. There is an easier way. Gather the mane into one long plait. This looks neat and tidy and is simply achieved. It will not work or lie flat with a short pulled mane; the mane needs to be at least 10-12 inches long and thick enough to give the finished plait weight and body. Again the diagrams show how the plait is done. The

forelock can either be plaited or not, depending on which suits
your horse's head best. This plait should not be used if your
youngster has a thick or poor-shaped neck as it will tend to
emphasise the fault. Thoroughbreds with thin silky manes do
not look well plaited this way either, but it suits pure Arabs and
native ponies to perfection.

Not that any of these breeds should ever be shown plaited,
but long unwieldy manes often need to be tidied away for
ridden events to prevent the rider's hands becoming tangled
up in the mane. I have tried to pull up our senior stallion from a
canter by his own mane having lost the requisite string in a
muddle of hair! A long mane flying into one's face on a wet day
can cut and sting and again the long plait will take care of this
problem.

Native pony breeds and pure-bred Arabians are always
shown in their breed classes, whether in-hand or under saddle,
completely unplaited. A judge in a native pony class is entitled
to ask a plaited exhibit to leave the ring, and of course the pure
Arabian's flowing silky mane and tail are his crowning glory. It
goes without saying that when manes and tails are not plaited,
even more care must be given to washing and fingering out, so
that each hair hangs free and untangled.

One other thing needs practice before the show day: boxing.
Everyone knows how to box a horse, but one still hears of
animals being bad to load and of accidents occurring, so a few
hints may not come amiss when you are setting out to box a
youngster, or mare and foal, for the first time.

I always prefer to convey a young horse or a mare and foal in
a horse box rather than a trailer. Trailers are claustrophobic and
horses find it difficult to balance themselves because they travel
facing forwards. If one thinks about it, the reasons are
obvious: a horse coming to an abrupt stop out in the field pulls
up with its weight on its quarters, head up and hind legs tucked
well underneath its body. In a trailer, the animal's weight is
thrown forward every time the brakes are applied or there is a
change of gear, and this means that he is trying to balance
himself in a way entirely contrary to nature. It is a fact that a
horse travelling loose in a container will *always* turn to face the

direction from which he has come—towards the rear. Food for thought with a young horse that has never done any travelling before!

A client of ours used to have terrible problems travelling an Appaloosa mare he owned. She was such a bad traveller that they could not exceed 5mph and had to keep stopping even on the short journey from our stud to her own farm, a distance of only about three miles. She always arrived at her destination in a muck sweat, and on several occasions she was walked home after service so as to avoid upsetting her. In desperation, as it was a time-consuming walk, my husband suggested loading the mare in the trailer facing the back—ie backing her in and allowing her to look out over the ramp. A ring was put in the side of the trailer to which she could be tied, and from that moment all problems ceased and she and her foal travelled to several shows, coming home with a rosette each time.

Before the horse is introduced to the scene, position the box so that there is a hedge or wall against one side of it, and extending back past the ramp. If you have a narrow lane with hedges on both sides that is even better. Lower the ramp and make sure that it does not move or rattle. There is nothing more disconcerting for a youngster than a ramp that booms and shakes when he puts his foot on it. After this lead your youngster out of his stable, up the ramp and into the box. Lots of time, lots of patience, no sticks, no ropes, no shouting. Just patience, a scoop of oats and a sympathetic helper or two who can lift feet and place them on the ramp as required. We have never failed using this method, the only exception being with two older horses, both of whom had been previously frightened and would start to fight the moment they saw a box.

With previously scared animals like this, there is one infallible method. You need three people and the box parked as described earlier, with a hedge on both sides of the ramp. You also need a sheep hurdle, a gate or, even better, a piece of thick plywood 6 x 4ft. It is essential that it does not bend. One handler leads the horse firmly towards the box, the other two assistants follow up the rear, holding the sheet of plywood or the hurdle vertically between them. Should the animal hesitate the plywood is brought into contact with his quarters, kept there and indeed moved forward, the leader also moving forwards but not pulling the horse along or looking back. Once the recalcitrant beastie feels something he thinks is solid up behind him, he will move forward, and the only place he can go is up the ramp. Once in the box the horse is then patted and rewarded. Please notice that again there is no shouting or beating, just the old principle of determination and patience. I would add that the two awkward customers were cured after being loaded in this way a time or two.

Never, ever allow either the horse or yourself to get ruffled or irritated. That is a sure-fire way for accidents to occur. Make certain that you have a long lead rein on the headcollar. There is nothing more painful than having a headcollar rope tweaked through your fingers, and also it is far easier for a youngster to get loose should he run back suddenly if the lead rope is too short.

If a young horse is boxed with patience and sympathy a time or two, he will quickly become confident and load speedily and quietly without fuss. If you have a quiet companion whom your youngster knows (perhaps his walking escort), it is not a bad idea to take him along for the ride and to act as a confidence-giver and a lead if necessary.

I think I have seen more fiascos at shows caused by people making a muddle of loading a mare and foal than from any other cause. If you are travelling a mare and foal, the foal must be loaded first. Two handlers adopt the baby-walker position and propel him up the ramp. The third person follows immediately behind with the mare. Do not be tempted to load the mare first, hoping that the foal will follow: he won't, and the mare may well panic when she gets inside the box and finds that her baby is not there. If she swings round in her anxiety she may well flatten her leader against the side of the box with the risk of broken wrists and cracked ribs. The foal also is likely to panic if mum rushes off into that big black hole without him. If however the foal is put up first and held close (still in the baby-walker position) by two confident handlers, he will not become too alarmed and the mare will have an added incentive to go up the ramp. It is essential that a mare and foal travel loose, with room to move about. Headcollars are of course left on both of them. (A youngster other than a foal can travel tied up if necessary, providing he has been taught to tie up first.)

One last hint on the subject of travelling. The floors of some containers and trailers can become exceedingly slippery with use. A little sand scattered on the floor before the straw is put down will solve the problem.

Two days before the show your exhibit should be bathed if necessary. The day before, clean and prepare all your equipment, including your own clothing, give your horse a final grooming (all bar the last polish) and hope he doesn't lie in a dung-pat overnight. Grey horses always do! I frequently plait manes and tails late in the evening before an early show. It does save time and temper in the morning and it is usually quite successful if you stitch your plaits in. Your horse's tail should be plaited and put in an old stocking which is then bandaged on. This will keep it clean. Do not remove the bandage until just before you go in the ring. White socks can be finished off with a chalk block for extra sparkle, and of course hooves should be oiled, both things being done just before your class.

Try to set off in good time so that you allow yourself about an hour before your class in which to collect your numbers, and ascertain the whereabouts of your ring. Having found out how late the classes in your ring are running, delay unboxing your exhibit until your class has actually been called into the collecting ring, as every animal shows itself much better if it still has the fresh sparkle. Make your way to the collecting ring. Remember your number and to remove your tail bandage. Tell the steward you are there and . . . good luck.

At the end of the day don't forget, as you eat your fish and chips on the way home, that your horse would probably like a nice bran mash. Don't go into a restaurant to have a celebration meal leaving a tired horse outside in the box to wait and get cold.

12
A Riding Horse at Last

Your youngster is three this year: time at last to climb on to his back and ride away, king of all you survey, on your own home-bred charger. If only it were as easy as that!

The foundations of your colt's education will have been well and truly laid during his first three years and by his fourth summer he will be physically strong enough to carry a rider on his back for short periods. His approaching mental maturity will also allow rather more rapid progress to be made with his training. By the autumn of his fourth year he ought to be capable of being ridden at a walk, trot and lolloping canter in company, and for short periods alone, and generally be on the way to becoming a self-possessed young horse.

It is my firm belief that all young horses should be professionally backed, being sent for about three weeks to a well-recommended trainer to be lunged, long-reined and sat on. The person chosen must be someone whose reputation in the handling of young horses is beyond reproach. Again I cannot stress too strongly that you must visit the establishment you have selected and see for yourself how they work. Are they quiet and sympathetic with the youngsters in their charge? Do they listen to you when you tell them how much training your colt has already had? Or do they just say, 'Oh yes, well of course amateurs always say they have done things and they never have. We always start right from the beginning.' Personally this attitude makes me see red; if, for instance, a colt is already mouthed and accustomed to a roller, is it not better that the trainer is told?

Avoid such places; avoid also the rough, the brutal and the plain ignorant and inefficient. Better to muddle along alone, taking things gently and slowly, rather than risk someone frightening your carefully reared colt and shattering his confidence in himself and humans in general.

The amateur, however knowledgeable, often lacks the time, facilities, equipment and confidence to back a youngster efficiently. But having written that, I sat and thought about the establishments available in our own district, and I had to admit that with one or two notable exceptions there was nowhere I would have happily sent a youngster of mine to be backed.

So for those of you who feel confident enough to back your own youngster, I will describe the method we use, which is unorthodox but has proved simple and effective. The main attribute needed by anyone considering tackling this job is quiet confidence. If you are the nervous, highly strung type who is always worrying about what is going to happen next, backing colts is not for you. You will only communicate your nerves to the young horse and make him apprehensive and jumpy.

It does not matter if you are not a polished horseman. The essentials are an independent seat, quiet hands, and legs and calves that stay still and do not clutch. Common sense will sort out many problems before they ever become problems at all, and prevent you from making avoidable mistakes such as trying to put on a crackling nylon anorak because it has suddenly started to rain when you are on a colt that was only backed for the first time five days previously. No, it was not us and yes, someone was thoughtless enough to do it, and yes, the horse took off, the rider fell off and then sold the horse as unreliable!

My method is not the classic lunge, long rein, back, routine. Most small establishments do not have adequate facilities to lunge a horse properly, and no amateur should ever attempt to long rein a horse; the results are usually disastrous. The backing technique I use is simple, effective and does not need the use of a covered school or expensive equipment. We will come to the details later.

Your youngster's three-year-old training should start in the early part of his fourth summer, once the weather has become reasonably settled and he is living out at grass. It is not a good idea to try to back a stabled corn-fed youngster with all the gales of March blowing through his tail. Start off by re-introducing the bit, bridle and roller. These things should make a gradual reappearance, but there is no need to go back to the stage of tying the bit on to the headcollar. Your youngster should be walked out in hand for about a week, being led from reins attached to the bit, with loose side reins fitted. A full set of light shoes should be put on too; a young horse cannot work happily with sore feet. After about a week the roller should be exchanged for an old light saddle without irons or leathers. Introduce it very cautiously, allowing the colt to have a good look at it and sniff it before you attempt to lower it on to his back. Make sure that the lining of the saddle is warm—stand the saddle facing the sun for an hour or so first, or put it on another horse until such time as you need it.

You must have an assistant to place the saddle in position. Your place is at your youngster's head to reassure him and keep him still. You cannot possibly restrain a horse effectively and also put a saddle in position yourself—you need both hands for both tasks! If the horse should move forward quickly when he feels the saddle touch his back, your helper can just lift the saddle clear, but if you are trying to manage alone and the colt moves suddenly, the saddle may well fall to the ground with a bang, giving him a fright it will take him a long time to forget.

You do not need a girth on the saddle. For the first time or two, just place the saddle in position and lift it off again, repeating the exercise two or three times. The next step is to add the girth and *gradually* tighten it. Your colt can now go for his daily walk wearing his saddle instead of the roller.

Side reins should still be used but they must be attached to the saddle by a flexible link. While the colt is being accustomed to the saddle, the saddle seat should be patted and the flaps lifted so that he is really aware that he has something on his back and is accepting it—as distinct from not quite taking in that there is anything there at all. However, having said that, please remember that all new items of tack and fresh activities must be introduced very carefully and slowly. Any sign of tension, nervousness, fear or resentment means you have been too hasty. Retreat a step or so and progress less rapidly. The irons and leathers can now be put in position on the saddle, in the run-up position of course. I always tie the leathers together under the animal's belly so that there is no danger that they will flap when he is trotted. A surcingle over the saddle will also eliminate any movement of the saddle flaps if your colt shows signs of apprehension. The next stage is to lower the stirrup irons so that the feel of them against the horse's sides does not worry him when the rider finally mounts. They are also tied together underneath the belly.

Whilst you are out on your daily walks, it is a good idea if your assistant can stand on the verge, bank or other convenient object and get the horse used to seeing someone above him. This must be done quietly and sensibly, no attempt being made to touch the horse or his equipment. He is just led alongside someone standing on something high enough for his or her head to be at a rider's height above the youngster's back. This exercise, I find, is of tremendous value in the quiet acceptance of a rider.

The time has now come to start the backing process in earnest. Position a stout wooden box about 24 inches high in a quiet corner of the yard, away from the horses' boxes, and out of sight of any that may be turned out in the field. It is essential that your young horse concentrates on you and on what is going on, not on any distractions around him.

When you return from your walking exercise, lead your colt alongside the box, letting him inspect it first if he is at all suspicious. Make him stand beside the box. Your assistant, the person who will actually get on to the youngster for the first time, gets on to the box and gently presses down on the saddle with both hands. Repeat the process, and if all goes well the assistant can then lean his full weight across the saddle, but still keeping his feet firmly on the box. Over a period of about a week the rider's full weight can be transferred to the colt's saddle, until such time as the rider is lying across the seat of the saddle with feet off the box. I repeat, and cannot stress too strongly, that progress towards this point must be slow and sure. At any sign of tension or unease, retrace your steps slightly and go more slowly. At this stage no mistake is allowable.

The old saddle should be exchanged for a saddle that fits both horse and rider before the colt is backed.

There are three ways of actually getting someone astride the colt's back for the first time. The rider can lean across the saddle as usual, and if the animal remains relaxed just quietly swing a leg across, being careful not to drag a toe across the quarters or dig him in the ribs, and then cautiously sit up. If the colt is accustomed to weight being placed in one stirrup it is perfectly possible to mount by the stirrup in the normal way, but this method has one big disadvantage with a young horse. Should anything go slightly wrong whilst the rider is half up, he is stuck with one foot in the stirrup and cannot abandon ship with the requisite speed, ease and minimum of fuss. If you do decide to mount by the stirrup, there must be a third assistant on the off side to hold the off-side stirrup and prevent the saddle slipping round.

The most usual method of getting the rider aboard is for the third assistant quietly to leg the rider up. Everyone knows how to give a leg up, so I won't go into details, but do make sure both the rider and the legger-up know what to do.

On the day you choose for backing your colt, try to make certain that no untoward occurrence can disturb the peace of the stable yard. Pick a dry, warm day, wait until the postman and the baker, etc, have called, shut the dog indoors, and take the telephone off the hook. A third assistant is useful to hold the off-side stirrup and generally steady the colt, even if not required to give a leg up, but make sure that the extra person is as quiet and calm as you are. The colt's handler stays at his head. It is essential that the human he knows best is there to give the orders and reassure him. That is why the handler should not be the backer. Actually, a light and supple teenager who is quiet and confident is possibly the best person to sit on a colt for the first time.

Take your youngster for his customary exercise, and on return lead him up to the mounting block as usual. Your rider, hard hat on, of course, leans across the colt's back, and then if all goes well mounts by whichever method you have decided to use. The rider must sit very gently, and must not arrive in the saddle with a great flop. He should sit up even more gently and quietly, stay there for a very few minutes and then equally quietly get off again. That's it. Your home-bred horse has carried a rider for the first time.

In 99 cases out of 100, backing is as simple and unexciting as that, always providing that the preliminary work has been slowly and carefully done and the rider is quiet and confident. How often I seem to use these words, but they are the key to success with young horses. If you think a colt is going to do something sinful, he probably will. Likewise, all animals can smell fear, so if you or your rider are really nervous it is no use your trying to back your colt; you will only upset him when he smells nerves and gets worried wondering why you are worried. Fear is highly contagious, and especially so amongst animals whose natural protective and escape mechanism is flight.

The backing process is repeated for two or three days, and then providing the horse remains relaxed the handler can lead him forward a few strides, no more. The rider at this stage remains passive, does not pick up the reins and does not put more than a toe in the stirrups—just sufficient to stop them banging about. The rider's hands should be firmly in a neck strap and she does not attempt to use her legs.

If there is any sign of tension on the colt's part, the rider will be the first to notice it—possibly a stiffening of the back muscles or sudden movement of the ears. A quiet 'whoa' from the rider should be sufficient to make the handler bring the colt to a halt at once. If the tension persists, the rider quickly dismounts—both feet out of the stirrups and a quick jump to safety. It is essential at this stage of the proceedings not to have an upset of any kind. Whatever happens, *you do not fall off a young horse;* the fright received will be remembered long after the incident that caused the trouble is forgotten.

As the days pass, the distance ridden can be gradually increased, until the rider is getting up as soon as the colt is brought out of the stable, dispensing with the walking exercise first, excepting just a circuit or two of the stable yard to ensure that his saddle is really warm and that he is settled. Once this stage is reached it is obviously vitally important that the horse is tacked up at least half an hour before he is needed so that the saddle is thoroughly warm before it is sat on. Failure to observe this simple precaution has resulted in more unnecessary bucking and other upsets than any other single cause. The regular use of a numnah is a good insurance against a cold back too.

If the rider has not so far been mounting by means of the stirrup, this should now be introduced. Care must be taken that the off-side stirrup is steadied, so that there is no fear of the saddle being pulled round as the rider gets up. Too often a young horse returns from the breakers without the stirrup ever having been used at mounting time, the rider always having been legged-up. When the stirrups are pulled down prior to mounting do make certain that you do it gently so that there is not a sudden crack beside the colt.

As a teenager, I regularly took livery hunters to the meet for their owners. On one occasion I had in my charge two fit, clipped thoroughbred hunters, both of whom were being qualified for point-to-points and were consequently rather full of themselves. We arrived at the meet, and the unthinking owner of one of my charges rushed up to me, claimed his horse and before I had even had time to hand the reins over to him, pulled down the off-side leather with a tremendous crack. This was just too much for a corned-up thoroughbred on a cold morning; he jumped sideways and cannoned into my horse who, equally upset, put in an almighty fly-buck, jerking the other horse's reins out of my hand. His potential racehorse was eventually caught up some five miles down the road. Luckily roads were quieter then! I confess to having left the thoughtless owner to hunt his own horse whilst I went in pursuit of the more usual quarry and my hunter's certificate.

It is essential that the youngster is made to stand absolutely still while the rider gets up. Bad habits such as walking forward, whizzing around, etc, should not be tolerated and indeed should never be allowed to start. If the youngster's regular handler is always there when the rider gets up, and steadies the colt by his off-side rein and stirrup such restraint, combined with the fact that the rider is by now picking up the reins in the normal way as he mounts, should ensure that the youngster remains standing still. If he does move, do not scrabble aboard; return to the ground, make him stand and start again. Persist, until he does stand still until the rider's seat is in the saddle.

I have said before that it is essential for the rider to sit in the saddle quietly without a great flomp. If a young horse is unlucky enough to come across a rider who arrives in the plate with an unthinking wallop every time, then it will not be long before the youngster does show resentment in the shape of rushing off and bucking.

Once your young horse has accepted a rider and is relaxed about the proceedings, you can once again venture out on to the road. The handler still leads the youngster, but now from a lead rein attached to the noseband. The rider gradually becomes more active and takes over the direction-giving and progress-maintaining. The rider must be quiet, confident (those magic words again) and certain of his or her actions—in other words an independent horseman with a secure seat and quiet sensitive hands.

For the first weeks of the colt's career under saddle, the rider should emphasise his or her aids with the same words as the handler has always used, eg 'Walk on', 'Trot on', etc. As time passes, the handler's presence is merely a safety measure and the lead rein should be dispensed with as soon as it is safe to do so. It is not a bad idea for the rider to continue to carry it then if necessary it can be passed back to the handler in moments of stress.

It is essential that a young horse goes freely forward at all his paces and that he is taught from the beginning to adopt regular even paces. It is the rider's task to see that this happens. It is better to be satisfied with a short distance at an evenly paced trot than to try to keep it up for too long a distance so that the horse gets tired, loses concentration and starts slopping along or dragging his toes.

Do not make your rides too long at first. A quarter of an hour is quite long enough for the muscles of an immature horse to support the weight of a rider, and if the rider has been doing his or her job properly he will be tiring, too, at the end of the period. It has been truly said that riding a young horse for fifteen minutes is more tiring than the most exhausting day's hunting.

By the time your youngster has progressed to trotting regularly, the handler will probably be only too happy to continue escort duties on a quiet horse, preferably one that is known to the colt. Once both parties are mounted, progress can be rather more rapid. It must in any case be daily at this stage, if you want your colt to progress in his education. Little and often is the secret of success with a young horse. Daily take him a little further and ask a little more. By this I do not mean keep asking for new movements, but look for regular improvement in the colt's way of going. Activity, cadence, rhythm and impulsion should all improve as the colt's muscles become accustomed to carrying a rider.

A horse should always walk fast and trot slowly (not a sloppy crawl, but a swinging, on-going, steady pace.) One other thing: when you are trotting, remember to change the diagonal regularly. I always change every 100 strides, or between each telegraph post if I am trotting along a road.

If you are unsure how this is done, here is the method. A trot is a two-time pace, in which opposite pairs of legs move together, eg, off-hind or near-fore, then near-hind or off-fore. Obviously at a rising trot the rider's weight will be off the horse's back as one set of legs hits the ground and in the saddle as the opposite pair come to earth. If one always sits as the same pair of legs come down, the muscles of the horse's back adapt to the uneven load and become stronger on one side than the other, the horse so becoming one-sided and stiff. All one does to change diagonal is to sit for one stride only and then rise as before. You can check that you have in fact changed by noticing which front leg is moving forward as you sit.

Hill work is very helpful in teaching a young horse to balance himself. Try to do more uphill work than downhill until your youngster has developed at least a semblance of natural balance. This possibly sounds silly, but try to tackle any really steep hills in an upwards direction. If you give your ride routes a little forethought, you will usually find that it is possible to avoid any really severe downhills. Going down a steep hill is very difficult for a completely unbalanced youngster whose weight is all on his forehand, but going uphill he has to get his hind legs under him to push himself up, and this in turn will help develop the muscles of his back and quarters; that will make it easier for him to balance himself and get the weight off his forehand.

The rider must keep a light but positive contact with the youngster's mouth at all times. This contact must never vary, but equally importantly the hands must follow every movement of the horse's head and neck whilst retaining sensitivity and lightness.

In this way a colt will soon learn to trust his rider's hands and accept a contact. He will then feel safe to use this contact to balance himself. Let there be no mistake: a young horse will at times take quite a strong contact if the rider's hands are to be trusted, and this contact will only start to lighten as the horse starts to balance himself. If you stop to think about it, the reason why is obvious. When a rider first gets on to a young horse with its undeveloped muscles, the youngster will tend to raise his head to compensate for the weight on his back. You will realise this must be true if you think about what happens if a heavy rider is underhorsed: the animal, no matter how well schooled, immediately starts to go with its head in the air, hollow-backed and hocks trailing out behind.

As the youngster's muscles strengthen he will start to lower his head and then the head will be carried low with the weight on the forehand. This is the stage at which the rider's hands are vitally important. If they are steady, the youngster will take a somewhat heavy contact and use his bit as a prop whilst he learns to bring his hind legs under and balance himself, lightening his forehand and shifting his centre of gravity back. Once this starts to happen, the contact with the horse's mouth will lighten as he learns to carry himself without relying on the rider to hold him up. His head carriage will become better too, but this is the last thing to come right.

You cannot force a horse's head and neck into an elegant shape until everything else is working properly. A vast number of young horses are frightened and have their mouths ruined by insensitive, ignorant riders trying to heave their heads into the imagined 'correct' position or putting all sorts of unnecessary ironmongery into their mouths because they are alleged to pull. These riders will not look at the problems objectively and realise that if only they helped the horse to balance himself the problem would go away.

This balancing process must be achieved through forward movement and may involve fairly vigorous use of the legs to keep the impulsion going. A horse only becomes forward-going and balanced by going forward into his bridle, and this he is not going to do if the rider's hands are always pulling at his mouth. Hands that haul, jiggle about or fly up in the air should not be in contact with a young horse's mouth.

Balance and free forward movement should develop together, and it is equally certain that you cannot have one without the other. It is equally certain that no horse is ready for further schooling until they have been achieved.

Our three-year-old's work should be directed solely towards this end. Never mind the circles, the trotting poles and the other so-called suppling exercises. Balance and free forward movement can be taught whilst hacking about the countryside without imposing on immature joints and tendons the strains and stresses of work on a circle. School exercises are, I am sure, a prime cause of splints and tendon injuries in young horses. Three-year-olds should never be asked to work on a circle. They are simply not physically strong enough. Their ridden work should consist of regular hacking over as wide a terrain as possible, uphill, downhill, walking and trotting, over ditches and logs, learning to open and shut gates, to cross water and enjoying an occasional lolloping canter on a nice gentle upward slope.

I always use a neckstrap when I am riding a young horse—either just a neckstrap or a standing martingale if the horse has any tendency to fling his head into the air, or to lighten his forehand to the extent of bringing his front feet off the ground no matter how slightly. The martingale is fitted loosely, of course, so that it does not act unless the animal's head gets much too high. It is far better to have a neckstrap to grab in moments of stress than risk using the reins to hold you in position. If, for instance, a youngster's unbalanced paces happen to shake you loose going up a steep hill, a hand in the mane serves the same purpose, but not all horses appreciate this. Holding the front arch of the saddle is no help to anyone; it does not get the rider's weight far enough forward to be of any assistance in taking the load off the horse's quarters.

When riding a young horse it is essential to carry a cane; a plain leather-covered one or even an ash plant. Long schooling whips or jockey flappers are not suitable and should never be used. The cane is used to reinforce one's leg aids if necessary and to correct the youngster if he does something seriously wrong such as offering to kick. When the cane is used to reinforce the leg aid it must always be used immediately behind the leg. Ask once with legs alone, gently; no response, ask again with the legs, strongly coupled with verbal encouragement; no response, ask again with legs and voice and reinforce your leg aid with a sharp tap from your cane—success! Do not flap the cane about indiscriminately and do not use it unless necessary. It is all too easy to get a horse stick-shy.

If a young horse really misbehaves, by offering to kick or some equally serious misdemeanour, he must be punished instantly and hard whilst being soundly scolded with your voice. Horses that kick are lethal and yet it is amazing how many riders just say a bland sorry without ever trying to correct their mount. If your horse kicks or offers to kick he must be rewarded with what our grandfathers would have called four of the best. Don't tap him: hit him, as hard as you can three or four times and shout at him at the same time, and start the punishment at the moment the deed is done. Drive him forwards vigorously with your legs at the same time too. Your youngster will quickly realise that what he did was not approved of and should not be done again. The only serious misdemeanour for which I do not hit a horse is bucking. I find that rather than giving tit-for-tat you are likely to get buck for hit, and anyway bucking is usually just a burst of joie-de-vivre—a lack of balance, rarely a vice. Horses who buck for a hobby are rare. In my opinion growling at the culprit whilst holding the hands up to get his head up and driving vigorously forward is a better remedy. Remember a moving horse cannot buck; he has to stop and put two feet together to get off the ground. A young un-balanced horse will often put in a buck if he is cantered down a slight slope. They get their weight on their forehand, lose their balance, get worried, semi-stop and put in a quick hump to get their centre of gravity back where it should be.

If a youngster suddenly finds himself in a large open space, excitement may go to his head and set off a bronco display of joie-de-vivre bucking. For this reason we are always very careful when venturing into fields or out on to open moorland with a recently backed youngster. It is a good idea on these occasions if the escort just takes the lead rein, as it does give an added measure of control. Try to tackle a large open space going uphill if possible, keep your horse moving on, keep his head up and your seat firmly in the saddle, even feet stuck for-ward slightly if you think danger looms! Insecure riders who perch over their horse's withers, leaning forward with a sickly grin, are asking to get parked. On a youngster, if in doubt, sit up, sit back, bottom well down in the saddle, heels down, hands up rather than down. You will be surprised how many juvenile acrobatics you can survive, and let there be no doubt about it, all youngsters will try their luck at some time, usually when you can least do with it.

The secret of being a good nagsman, and that is a good old-fashioned term applied to those rare birds who make it their business to bring on and train young horses, is to prevent situations occurring that can get out of hand. Again, if an argument with your young horse does develop, make sure as far as you possibly can that it happens when you are in a position to win. Don't, however, ride around expecting your colt to be wicked—such uneasiness communicates itself to the horse and makes him nervous and unpredictable. Ride your colt expecting to enjoy yourself, and encouraging him to enjoy himself as well. Your daily rides should be a pleasure for both of you.

Young horses can sometimes become nervous of strange sights or strange places. Your voice, used in a soothing, encouraging tone can do much to help him overcome his fears. I talk to a young horse all the time, about all sorts of things; the scenery, where we are going and anything that comes into my head. I get some funny looks from passers-by but youngsters usually go kindly for me.

Two more problems that may cause some concern are spooking at horrors in the hedge, and, growing out of that habit, whipping round. Of the two the latter is the more unpleasant and unseating, and if it is allowed to continue unchecked can lead to·rearing and nappiness.

If your horse does take an aversion to a strange object in the hedge, there is a right and a wrong way of coping with the problem. The first thing to remember is that you do not get off and lead him past the 'horror'—if you get off, you are at an instant disadvantage. You only have a short pair of reins to lead him with, he is much bigger and stronger than you; you are asking to be knocked over and possibly hurt, your horse stands a good chance of getting loose and being injured or killed, and above all else it is not doing his education as a riding horse any good. Riding horses are ridden, not led. Pass the lead rein to the escort if you are really worried, but stay on top. Using just sufficient contact with his mouth to prevent him whipping round, sit down and drive him forwards with vigorous leg aids, making soothing, encouraging noises. Keep your seat firmly in the saddle, sit up and kick whilst your hands say 'go forward'; don't grab hold of his head. The intention is to get him past the problem, not to stop him in his tracks. Don't turn his head towards the object; talk calmly and ride by as if the object didn't exist. If your colt really does get frightened, don't let the situation develop into a battle that could get out of hand. The escort should quickly sum up the situation and when it becomes obvious that the rider is in trouble, he or she should ride quickly between the horror and the young horse, hiding the problem from view and giving the colt a lead at the same time.

If your horse does succeed in whipping round, it is essential that he is turned back to face the object again and not allowed to complete a circle. He must be turned back the way he came and then driven strongly forward with the aid of the stick if necessary.

Actually there is quite an art in escorting a young horse. It is not a job for an inexperienced rider. The escort must always ride with great consideration, taking care to think of the likely effect of any action on the youngster. A competent and alert escort can do a tremendous amount to ensure that a three-year-old goes through the first months of his career under saddle without becoming frightened or upset. The escort should not however always act as lead horse; that is not the idea at all. The escort should be ridden alongside the youngster at first, only giving a lead when it is absolutely necessary, and as the young horse gains confidence he should be encouraged to stride along in front, laying the foundations for going out to work alone.

There has been rather a lot here about problems, but it is most unlikely that you will meet all of them; most likely you will not encounter any of them if you do things slowly and carefully and in a thinking manner. But it is often the un-corrected misdemeanour that leads to bad habits and a bad habit can become a confirmed vice and a real problem. Sorting out minor difficulties seems to be the thing that worries an inexperienced nagsman the most. The cry is 'He is so young, I didn't like to hit him!' If he has been *really* bad, then you must not be afraid to administer an instant reprimand.

The only thing remaining is to accustom your colt to having you adjust his tack while on his back. By degrees get him used to your tightening his girth whilst you are sitting on him, also shortening and lengthening your leathers. For the first time or two get someone to hold his head and murmur sweet nothings in his ear, as the raising of the saddle flap may cause anxiety if he catches sight of it suddenly out of the corner of an eye. Some years ago we had a youngster sent to us for schooling on. He had been backed as a three-year-old by a very well-known horse-breaker who had produced an impeccably mannered young horse. He was sent to us as a four-year-old to be ridden on for three months. There were no problems. The horse started to get fit and in so doing lost a little weight, with the conse-quence that his girth needed tightening a hole. Going down a rather steep hill, his ears seemed to be getting closer. I pulled him up and without giving the matter a thought raised the saddle flap prior to pulling up his girth a hole or so. The next thing I knew was that I was in full flight down the hill: nobody had ever before adjusted the saddlery while on his back. I also later discovered that the rider had never touched him from on top, other than to pat his neck. Coat tails and riding macs can be another hazard. Many a colt has discovered an unsuspected aerobatic ability when a riding mac has suddenly flapped across his quarters. With riding macs and girth tightenings, as with all other things, make haste slowly, and make sure your youngster becomes accustomed to all things peculiar without becoming alarmed or frightened.

Most of our horses will quite happily wear fancy dress or carry parcels, and I put it down to the fact that we have always introduced strange sights, sounds, equipment and activities with the utmost care and many kind words, beating an instant retreat at any sign of nervousness or tension.

If your colt has been ridden for about three months and is going sensibly and freely forward with regular easy paces, is quiet to mount, will pass most horrors and traffic, will go happily in front or behind and for short distances on his own, will open and shut a gate, cross water and go up or down hill without undue difficulty, you should be well satisfied. Your home-made horse is established on the road to success and has achieved as much as anyone should ask of a three-year-old. The moment has come to take his shoes off and turn him away for six months to finish growing. It is amazing how much stronger and bigger he will feel when he comes back into work at four years old.

13
Riding On

The basic schooling of a horse is a big subject and there are specialist books available. Nonetheless I have been riding and coping with young horses for more years than I care to remember, and my experience during that time leads me to believe that there are some points that many books overlook and are worth enlarging upon. Some of the hints will also be found helpful if you ever have a difficult or problem horse.

'Riding on' is the process of turning your green youngster into a tolerably well-schooled riding horse ready to progress to the more specialised training demanded by any branch of equestrian activity, such as dressage, eventing, show jumping or long-distance riding. The technicalities of improving balance and paces by means of basic schooling have been described by all the great equestrian authorities. Help can also be obtained by consulting some of the very excellent instructors who attend schooling sessions at riding clubs, etc. A word of warning though: do be sure in your own mind that the instructor really does know more than you, and also beware the slave of current fashion. Any instructor who says that, for example, every horse *must* be ridden in a jointed snaffle and a drop noseband, or that every youngster must do an hour's lungeing every day before being ridden, or makes any similar dogmatic statement with no regard to the make, shape or temperament of the animal should not be taken seriously. Many, many good horses have been spoilt in this way.

When you turned your three-year-old away at the end of his fourth summer, after being backed, he was still immature, possibly with quite a lot of growing and making-up to do. It is maturity or lack of it that should decide when your youngster is brought into regular work and his basic training completed. If he was a well-grown three-year-old, fairly mature in physique and with little more growing still to do, a break of about three months in the autumn of his fourth summer may well be sufficient rest for him to recover from the physical and mental strain of being backed. If, however, he was a great big gangling baby, higher behind than in front and with funny lolloping paces, he will benefit from being turned away for a full twelve months before being brought back into work in the autumn of his fifth summer, when he is in fact rising five. I hear cries of horror: 'What, keep him for nearly five years before I can do anything with him?' But if you wait until your horse is mature, or nearly so, his muscles and bones will have finished growing and he will be capable of carrying a rider and enjoying his work without risk of strain or injury, and what's more, the chances are that, barring accidents, he will still be enjoying his work twenty years later. Don't rush a young horse into work if you want him to have a long and trouble-free career under saddle. Thoroughbred sale rings are full of broken-down cast-offs who have been ruined by being ridden and raced at two years old, and we have all seen adverts offering three-year-olds for sale who are jumping coloured fences and hunting regularly.

Your youngster should be brought back into work gradually and thoughtfully. When he first comes in from grass, he should be wormed and shod, and during the first week or so reintroduced to a saddle and bridle and a rider on his back. There should not be any problems, but remember to proceed with caution, being certain that he remains calm and relaxed. Try to ride him for at least an hour every day with the exception of one rest day a week. On that day he should be turned out to graze, so long as the weather remains reasonable. A day at leisure, in the liberty of his own field, gives him an opportunity to have a buck, a kick and a good roll without risk to himself or his rider. I am convinced that a day's freedom occasionally acts as a safety valve and does much to prevent those untoward explosions that one can well manage without when training a young horse.

Do remember that when your youngster comes in from grass he will be totally unfit, and so for the first three weeks he should be kept to walking exercise only. It is no use expecting an unfit animal to improve his way of going, and to absorb basic schooling, if his muscles are not in a fit condition to cope with the work they are being asked to do. This is a sure way to stiffness and resistance. Fitness and education must proceed apace and the best basis for fitness is found in walking exercise.

This muscle-building period should also be used to improve your horse's walk. All horses should walk freely forward with a long easy stride that covers plenty of ground. The walk should be as fast as possible without loss of balance or jogging. My pet aversion is a horse that does not know how to walk and teeters along with little mincing steps covering no ground. All properly trained horses can produce a reasonable walk; obviously some are better walkers than others, but all horses using themselves properly should track up—that is, the track of the hind feet should overlap the track of the front feet by as much as 6 inches. This is a good guide to the amount of effort that your youngster is making. Remember that all impulsion and propulsion comes from the horse's hindquarters and so your youngster must be encouraged by sensitive but active use of your legs to bring his hind legs well under him and use the impulsion so gained to swing himself along. This long swinging stride is aided by the pendulum effect of the head and neck. If the rider's hands resist this movement by hanging on to the horse's mouth and keeping him on a short rein, you cannot expect to establish a good swinging walk. A horse that is really walking on will require nearly the full length of his reins, but of course a light but positive contact must still be maintained by the rider's hands following every movement of the horse's head and neck. After a while, your horse should be able to sustain a quick active walk for long periods without chivvying; in fact his walk should always be the quick, long-striding variety. If he is always made to walk this way at the beginning of his training, it will become a matter of habit and be a tremendous asset to him. There is a lot of truth in the old saying that any fool can make a horse gallop but it takes a horseman to make him walk.

The thing that will strike you once you have your youngster back into work and just starting to get fit is how much stronger, bigger and more powerful he feels after his rest. It is amazing how young horses change at this stage of their training; suddenly they are gangling babies no longer. Muscles begin to form shape and outline alters for the better as the horse's natural balance improves, enabling him to carry himself better.

The change of shape occasioned by maturity combined with increasing physical fitness can lead to problems with the fitting of a youngster's saddle. This is a point that is often overlooked. A saddle that fitted perfectly well when he was first backed will almost certainly need the stuffing adjusted as he matures. If you are unlucky you may even need a new saddle; it is obviously essential that if your horse is to go kindly and happily for you, and enjoy his work, his tack fits perfectly and does not cause any discomfort. Remember, too, not to fasten his throatlash so tightly that it nearly throttles him if he tries to drop his head and flex at the poll. A correctly fitted throatlash should allow room for a clenched fist alongside the jaw bone. The throatlash has no function other than to prevent the bridle coming off over the horse's ears should you leave the saddle somewhat suddenly by the front door! The girth too should admit two or three finger-widths when pulled up sufficiently to keep the saddle in place. If it needs to be tighter than this, then the saddle does not fit or your horse is too fat. Either way the remedy is obvious. To use an over-tight girth is the best way I know of stopping a horse; after all, you don't try to run a mile in a pair of tight corsets. Our senior stallion will not come out of his stable if his girth is too tight; he just stands in the middle of the box and glowers until it is slackened off a hole.

After your youngster has been back in work for about three weeks, the worst of the grass podge should have disappeared and some slow trotting work can be introduced. The amount of trotting can be increased as time passes, as can the length of the exercise time and the distance covered. If he has been brought back into work in the autumn and it is intended to keep him stabled and working throughout the winter, now is a good moment to think about removing a little of his coat so that he does not sweat up and lose condition as you introduce some faster work. My first clip on a young horse is a very simple affair: from throat to chest, between the front legs, and just clear a little space over each shoulder. The diagram shows what I mean. It is surprising what a difference this exercising clip makes to a horse's comfort. A full trace clip and a hunter clip can follow later, when the horse has grown his full winter woollies and is rugged up at night. Much condition can be lost and a young horse can easily catch a chill if the creation of this breathing space is delayed for long after the animal has started any significant amount of faster work.

When you start trotting, aim for a pace that is regular, even and not too fast. A free-flowing rhythmic stride with a good cadence is the ideal. One should not attempt to ask for any collection at this stage of a horse's education. All work should be done on a longish rein, with a light, even and giving contact

Hatched areas show a first clip suitable for a young horse just into work

maintained by sympathetic hands. Collection is, in essence, energy compressed into brilliance and sparkle. If energy in the shape of free forward movement is lacking, there is nothing to compress into brilliance. Therefore, attempted collection will only result in roughness and resistance.

As well as working to improve your youngster's paces, it is essential to work seriously to get him properly fit for the first time in his life. Three weeks' walking exercise will have laid the foundations for fitness and now alternating periods of walking and trotting should be introduced—what I call Scout's pace. Scout's pace consists of walking for say 100 paces, trotting for 100 paces, walking for 100 paces, and so on. As your horse's physical fitness improves, the ratio of trotting to walking can be increased until one is trotting for say 400 paces to 100 paces of walking If you are riding along the road, telegraph poles make very useful markers. This form of exercise has the advantage that it increases the demands made on the horse's physical resources without any unnecessary stress, the walking periods always giving him time to catch his breath and cool down.

The secret of getting a horse fit is to apply just sufficient stress to increase the capacity for work of his heart, lungs and muscles by a fraction every time he is ridden. Fitness, as I am sure everyone knows, is not achieved by riding a horse until he is exhausted. I always try to pull up for a breather at the walk just before the horse starts to sweat. If you watch the base of your horse's ears, and as soon as they become a little bit damp pull up for 100 yards' walk, you will prevent him getting really 'sweated up'. A horse should not lose weight with increasing fitness, unless he was grossly overweight to start with. A fit horse should not resemble an animated clothes-airer: he should be big, round and hard with packed muscle, particularly on his neck, shoulders, quarters and the inside of his thighs. A horse in regular work should always receive two corn feeds a day, even if he is living out, and his concentrate ration must be increased in relation to the amount of exercise.

One little word of warning though: this increased fitness, combined with the feeling of well-being it induces, may just go to his head, so beware the unexpected hump or spook. There is only one thing more unpredictable than a green horse and that is a half-fit green horse! Should your youngster get slightly above himself, remember that there is nothing like a long steady trot, preferably up a long steep hill, for taking the steam out of the situation. It will get him even fitter too.

By this time of course your horse will be doing some cantering work and possibly one or two sessions of basic schooling on the flat every week. Try to introduce as much variety into the activities as possible; alternate hacking with schooling sessions, try to vary your route so that you don't go the same way every time you go out; introduce diversions by jumping any little logs or ditches you come across—do any little thing to keep him interested, whilst all the time improving his way of going, fitness and general knowledge of the world.

A word of caution, though. Confine your jumping to little obstacles when you are riding alone. Never try to tackle a sizeable fence on any horse, however well-schooled, if there is no one with you. The best horse can make a mistake, the most capable horseman can fall off, and should either horse or rider be injured away from help, who is to know and who is to go for assistance? Make it a golden rule never to jump any significant obstacle unless there is someone with you. I personally know of two very nasty accidents that were caused by folks jumping big fences alone. One man on his way home from a dull day's hunting decided to give his mount a school over a big timber fence; the horse caught a toe, turned a somersault and broke his neck. The rider was pinned to the ground under his dead horse and was found in a desperate condition the following morning, the night having been bitterly cold and wet. In the other incident a point-to-point rider decided to have a jolly round a friend's cross-country course; his young horse refused, at speed, posting his rider over his head and high-tailing it for home. The rider severely injured his back in the fall, but in his semi-concussed state did not realise just how much damage he had done to himself and managed to drag himself home. He was later told that had the damaged bones in his back moved just a fraction of an inch he would never have walked again.

Hunting is an excellent way of broadening a young horse's view of the world, and teaching him how to look after himself whilst finding his way across an unknown country. For the first time or two it is a good idea to select a less popular meet about four or five miles away from home, so that you can hack there and back and are unlikely to be trampled underfoot at the meet. When you arrive at the selected venue try to select a quiet corner away from any over-fresh animals possibly ridden by uncaring, gossiping folk, too busy nattering to pay attention to their horses. Let your youngster look about him and take in all the exciting sounds, sights and smells. When hounds move off, let him take his place and trot on with the rest of the field, taking care that he does not get barged or jostled—which would invite retaliation in the shape of kicking. Keep at least two lengths away from the horse in front—all horses kick, some kick more than others, and you do not want your nice young horse kicked on his first day out. At the first draw, again keep a little apart from the crowd. If you have no friend who can accompany you on a seasoned hunter to act as escort and lead to the young horse, try to select a seasoned hunter to follow; most people are only too happy to help educate a young horse if the youngster is being ridden sensibly and allowed to go on.

Do not stay out very long for the first time, and hack back quietly so that the blood has gone off the boil by the time you get home.

If, in spite of your precautions, your horse is still in a tizzy when he returns home, it is a good idea—provided that hounds are not within earshot—to take off his tack and turn him straight out into the field for a roll and a nibble of grass. Don't leave him out for more than an hour, but it is my experience that this routine will avert trouble with any horse that breaks out into a cold sweat after a little excitement or exertion. I do not know if it is the coating of mud that acts as insulation or the nibble of grass that relaxes over-tense muscles, but I do know that it works! It is much easier to put your neurotic hunter into the field for an hour (fifteen minutes is enough in midwinter) while you have a cup of tea than to spend all night changing rugs and walking a sweating horse about.

Incidentally, remember not to fuss a tired horse; just make sure that there is no mud that will chafe him under a rug, and leave the main grooming until the morning when he will have rested and recovered somewhat from his outing.

By the next time he sees hounds, your young horse will have learned that the sound of the horn and the sight and smell of hounds mean exciting happenings and he may become quite a handful. A huntsman friend of mine always said, 'I'll hunt a young one for the first time, but someone else can take him for the second and third time!' The first time he goes out with hounds he will be so busy looking about him that his behaviour will probably be quite circumspect, but the time or two after that may well be quite exciting for both horse and rider. He is quite likely to put in a buck or two if a big field suddenly goes rushing off at a gallop; he may find it necessary to stand on his hind legs to get a better view of proceedings (a standing martingale is a sensible precaution) and almost certainly it will not be possible for him to stand still! There is nothing for it but to ride your youngster on through this stage. As the old horse-men used to say, 'Ride through your troubles.' Stay calm, keeping your youngster going forward, avoiding getting crow-ded in gateways, and if there is any jumping, get his escort to give him a lead; go round the easier way if you think that any obstacle is beyond his ability. Don't overface him and risk giving him a fright. Do make sure that you turn his head towards hounds if they pass you in a lane and jam his quarters in the hedge so that there is no risk of him kicking out at a hound in his excitement.

After four or five outings with hounds, your youngster should have sobered up and become fairly sensible, always providing that he has been allowed to go on and not been so messed about at the back amongst the refusers, the no-hopers and the my-dear-isn't-it-exciting-I'm-scared-I'll-fall-offs that he has become frustrated and upset. As a child on my first day's hunting, I remember complaining bitterly that my pony was pulling and I could not stop. I was given a sound piece of advice that has stood me in good stead ever since: I was sharply told, 'Well of course he'll pull and mess about if all you do is hang about at the back. Go on after hounds and he'll behave sensibly.' I did, and he did! Much better to let your youngster stride on, well up with the leaders, for an hour and then take him home, rather than wander round the lanes all day achieving nothing except to get horse and rider cold and unsettled.

If your young horse becomes really over-excited and silly the first time he is introduced to hounds and hunting, or any other activity involving large numbers of other horses moving about at speed, it is no use saying, 'Oh dear, I don't think I'll do that again, he did behave badly!' The only cure is to keep going, and going again and again, until the novelty wears off and he learns to behave sensibly. It is helpful if you can hack to the venue so that the worst of the tickle is taken out of his toes before he arrives.

Hunting will do much to further the education of a young horse. It teaches him to cross a varying terrain with a minimum of effort, and to cope with the unexpected such as a ditch on the far side of a fence. Above all he should finally learn to gallop.

It is my experience that far too few horses are taught to stride on at speed in a balanced fashion. Of course, before a horse can be taught to gallop he must be fairly well balanced and on the bit, and also be reasonably fit. An unfit horse cannot be expected to go really fast and should not be asked to do so unless you wish to damage his heart and lungs. Once he is hunting and fairly fit, he should be encouraged to stride along if the opportunity for a good gallop arises. He should be picked up on to quite a short rein and encouraged to take quite a strong hold; do not get worried by this and think that you are being run away with. If you think about it, racehorses do not win races with flapping reins; there must be contact to aid the horse to balance himself. An unbalanced horse cannot gallop—he will sprawl about all over the place. At speed, a subtle balance is struck between the rider's weight and the contact with the horse's mouth, one helping to balance the other and vice versa. This is not to say that the rider should keep himself in position by hanging onto the reins; it is much less obvious than that. Action and reaction set to achieve perfect balance.

A horse that has been taught to take a decent contact at speed is much easier to stop than one that has been allowed to fall along anyhow. It does not take very long for a horse to realise that if the rider picks up the reins he is saying, 'Pay attention, we are going to go faster,' and so conversely it will only be a short time before the animal learns that the rider sitting up in the saddle and dropping the contact means 'Action over, steady up'. All our horses are trained this way from the beginning. Drop the contact means pull up. This technique will be found particularly useful with animals that have a tendency to warm up. If such are ridden on a loose rein and the sherbet antics ignored as far as possible, it is not usually too long before they start to go more calmly when there is nothing happening except gentle forward progress.

Actually I have found by experience that most animals who warm up, fizz and jog do so because they have never been taught or allowed to go forward. They are usually of a somewhat excitable temperament, and instead of the nervous energy being channelled into impulsion and activity, it comes out as 'hotness'.

There is one other general-purpose word of command that we always teach our young horses: that is some specific command that means 'Stop, look, think, don't mess about, this place could be dangerous.' We use 'Sensibly now'. You use whatever word you choose, whenever you meet a difficult place, and only then, and it is surprising how quickly your youngster will associate the command with potential danger and take the care necessary to negotiate the hazard successfully.

Some horsemen deprecate the use of the voice as an aid once a horse is over the just-broken stage. I do not; I firmly believe in its use to encourage, to calm or reprimand on occasion. By using the voice I do not mean uttering a constant stream of gibberish, but saying the odd word of encouragement or warning, and also passing on information. If, for example, you tell your horse in confident tones that you have already seen the polydragon in the hedge and emphasise your confidence with a tap with your stick on his ribs, it is more than likely that he will hastily realise that he was not really frightened after all and that there was no need to rush home to escape the Thing!

The other items of a youngster's education that often seem to be neglected are those little things that make him a generally handy, useful person to have about the place. All youngsters should be taught to open and shut a gate, go up to a letter box and post a letter, carry a shopping bag if necessary and lead another horse off. This is something that is not often taught these days but is incredibly useful if someone has, for example, to get off and hold an awkward gate open for others to pass: the spare horse can be taken on through the gate and out of the way. All these things are easily taught with a little care, remembering as with all things to proceed with caution and gradually so that your horse never gets a fright.

Just occasionally one runs into problems with a young horse's mouth. If he continually flings his head about, or carries it too high or yaws at your hands, obviously something in his mouth is causing him discomfort. The first and most commonly overlooked cause of mouth problems is tooth trouble. Get your vet to check that all is well. Perhaps he has a retained baby tooth or wolf teeth that need removing. If there is nothing amiss in that direction, consider your own hands: are you being too hard and unyielding, perhaps trying to pull your horse into the 'collected' shape you see in the dressage books long before he has developed sufficiently to be capable of more advanced work? When riding any horse, more particularly a young one, hands must be sympathetic, forward-going, quiet and sensitive to every movement of the horse's head and neck.

If you can honestly say that your hands are not the trouble (sometimes a candid observer who knows you well enough to speak frankly will have a different view), perhaps the problem stems from your choice of bit. I have had several animals brought to me for reschooling who were alleged to have mouth and head-carriage problems. All were being ridden in a jointed snaffle, and always this was the cause of the problem. Many horses, particularly Arabs and their crosses, have a slender, fine lower jaw which makes the nutcracker action of a jointed snaffle much more severe than it would be on a horse with a broader, coarser jaw. The effect is made worse if the bit used is perhaps a size too large. Many horses who do not go kindly in a jointed snaffle will go well in a straight bar snaffle, or if that does not provide sufficient braking power—and in most cases it will not —try an ordinary half-moon pelham, with a leather strap in place of a curb chain. If you do not wish to put your horse into a curb bit, and do not have sufficient control in a straight bar snaffle, the solution may be to use a bit like the Dr Bristol snaffle, which has the centre joint lengthened into a flat link with a join at either end, thus eliminating the nutcracker effect. With any mouth problem it is always worth experimenting with bits having different actions. Remember though to give the experiment time to work. Don't chop and change so often that the poor horse gets hopelessly confused and never has time to settle into any one bit that might turn out to be the one that suits him.

Just occasionally you come across the odd horse who goes much better without anything in his mouth at all, and for these horses a bitless bridle may be the answer. We have one mare who will never allow us to put anything in her mouth. She really resents a bit of any kind, but she goes well in a bitless bridle, which is really nothing more than a headcollar with reins. However, having said this, if you decide to use a bitless bridle, particularly the Hackamore variety with the reins on long cheek pieces, do make sure that it is correctly adjusted and used with care and discretion. Otherwise you are likely to inflict real and severe physical pain on your unfortunate horse. A bitless bridle in bad hands is an instrument of torture.

Whilst on the general subject of bits, mouth and hands, I was once witness to a fascinating demonstration at a jumping-instruction session at our local riding club. There were three horses on the ride that quite simply would not jump. They would not even approach a fence. The instructor patiently tried all the usual methods of persuading the recalcitrant animals to partake, all with singular lack of success. He eventually called all three to the centre of the school, removed the reins from their bits and attached them to their nosebands. Ignoring the cries of protest from the riders that they would not be able to control their mounts, the instructor sent the three guinea pigs out to trot a circle and canter on over the little fence that he had built specially for them: no lead horse, they were just to canter over the fence. All three animals promptly jumped the fence happily and willingly. What the instructor had spotted was the fact that all three riders had hard, unsympathetic hands which denied their mounts the head freedom that is necessary for happy, confident jumping. The lesson for those schooling young horses is, I hope, obvious.

On the general subject of tack and equipment, I must mention gadgets such as draw and balancing reins. In my opinion these contrivances are not only totally unnecessary but positively cruel. A horse's shape alters with exercise, maturity and physical fitness, and until his muscles are sufficiently developed he cannot use himself or begin to carry himself correctly. If by means of draw reins and other contraptions you heave the unfortunate animal into what strikes you as a desirable shape, undeveloped muscles are forced into positions they are not yet ready to adopt, and pain and stiffness must result, with subsequent resentment and evasion. No—get your horse fit, get him going forward with nice light even paces, encourage him to be supple by use of suitable simple exercises on the flat. Then you can ask for pretty outlines. You will be surprised at how easily they will be achieved without recourse to any artificial aid.

Several years ago I saw a most terrible act of cruelty. A horse belonging to one of our better-known show jumpers had been eliminated from a major jumping competition at a big show. The owner brought the unfortunate animal out of the ring and fixed a very tight draw rein from the girth through the bit rings and fastened it back behind the cantle of the saddle so placing the poor beast's head in an exaggeratedly low position, chin tucked nearly into his chest. She then proceeded to lunge him with his head fixed thus. I think the poor horse had probably been refusing because of back trouble, because he obviously couldn't use his quarters properly, and in a very few minutes sweat was streaming off him. He eventually stopped and refused to go forward at all, and when he was hit with the lunge whip managed a half-rear in desperation, lost his footing and rolled over. At this point his owner seemed to realise she'd done enough damage and led him off to the cattle wash, where the evidence of her cruelty was washed away. Be sure that if a horse is evading or resisting you, something is causing pain or discomfort. Do not try to overcome problems such as these with artificial contrivances. Think your difficulty through very carefully, with the help of your veterinary surgeon, if necessary, and see if a sympathetic horseman can't ride the problem out.

By the end of your youngster's first full year under saddle, he should have become a confident, happy, well-balanced ride, capable of turning his talents to any equine activity. Maybe you will take him into one or two little competitions, and perhaps you will be lucky enough to experience a thrill of legitimate pride as the public address system conveys the information that the winner was bred, trained and ridden by his owner. It is at such a moment that you know you made the right decision nearly six years ago when you decided to breed from the old mare, and that you have, more or less, been doing the right things ever since.

Index